Micro to Quantum Supercapacitor Devices

Supercapacitors have established their role as high-power density devices capable of storing energy for multiple cycles; these devices are more plentiful than batteries. This book outlines the fundamentals of charge-storage mechanisms in different configurations of supercapacitors. It describes the supercapacitor-related phenomena, state-of-the-art supercapacitor technologies, design and fabrication of electrodes, supercapacitor materials, macro-supercapacitor, planar supercapacitor, significance of electrode design, merits, demerits of current technologies, and future directions. It also details related physics, including prospective materials and electrode parameters.

Features:

- Provides understanding of the device architecture, electrode design, and pros-cons of classical supercapacitors
- Explains material design in the context of electrochemical energy storage
- Covers state-of-the-art quantum supercapacitor and technological challenges
- Describes advanced versions of supercapacitor devices, including macro-to-micro scale devices and applications at different scales
- Includes details of challenges and outlines of future designs

This book is aimed at researchers and professionals in electronics, electrochemistry, energy-storage engineering, chemical engineering, and materials science.

Emerging Materials and Technologies

Series Editor: Boris I. Kharissov

The *Emerging Materials and Technologies* series is devoted to highlighting publications centered on emerging advanced materials and novel technologies. Attention is paid to those newly discovered or applied materials with potential to solve pressing societal problems and improve quality of life, corresponding to environmental protection, medicine, communications, energy, transportation, advanced manufacturing, and related areas.

The series considers that, under current strong demands for energy, material, and cost savings, as well as heavy contamination problems and worldwide pandemic conditions, the area of emerging materials and related scalable technologies is a highly interdisciplinary field. This field needs researchers, professionals, and academics across the spectrum of engineering and technological disciplines. The main objective of this book series is to attract more attention to these materials and technologies and invite conversation among the international R&D community.

Nanotechnology Platforms for Antiviral Challenges
Fundamentals, Applications and Advances
Edited by Soney C George and Ann Rose Abraham

Carbon-Based Conductive Polymer Composites
Processing, Properties, and Applications in Flexible Strain Sensors
Dong Xiang

Nanocarbons
Preparation, Assessments, and Applications
Ashwini P. Alegaonkar and Prashant S. Alegaonkar

Emerging Applications of Carbon Nanotubes and Graphene
Edited by Bhanu Pratap Singh and Kiran M. Subhedar

Micro to Quantum Supercapacitor Devices
Abha Misra

Application of Numerical Methods in Civil Engineering Problems
M.S.H. Al-Furjan, M. Rabani Bidgoli, Reza Kolahchi, A. Farrokhian, and M.R. Bayati

For more information about this series, please visit: www.routledge.com/Emerging-Materials-and-Technologies/book-series/CRCEMT

Micro to Quantum Supercapacitor Devices

Abha Misra

CRC Press is an imprint of the
Taylor & Francis Group, an **informa** business

First edition published 2023
by CRC Press
6000 Broken Sound Parkway NW, Suite 300, Boca Raton, FL 33487-2742

and by CRC Press
4 Park Square, Milton Park, Abingdon, Oxon, OX14 4RN

CRC Press is an imprint of Taylor & Francis Group, LLC

ISBN: 978-1-032-00522-5 (hbk)
ISBN: 978-1-032-00523-2 (pbk)
ISBN: 978-1-003-17455-4 (ebk)

DOI: 10.1201/9781003174554

Typeset in Times
by MPS Limited, Dehradun

Contents

About the Author

Abha Misra is a Professor in the Department of Instrumentation and Applied Physics at Indian Institute of Science (IISc). She earned her PhD from the Indian Institute of Technology Bombay (IIT Bombay). She later received the Gordon and Betty Moore Foundation Postdoctoral Fellowship in California Institute of Technology (Caltech), USA. Dr. Misra is an Associate of Indian Academy of Sciences, member of National Science Academy, and TWAS Young Affiliate. She is also a recipient of INSA Medal for Young Scientists, SERB Women Excellence Award. Dr. Misra's research at IISc mainly focuses on meta-devices for self-powered energy storage and sensing applications.

1 Fundamentals of Supercapacitors

1.1 INTRODUCTION AND HISTORICAL OVERVIEW

1.1.1 Introduction

1.1.1.1 Basics of Electrostatic Capacitor

This section elaborates the concepts of supercapacitor in the context of differentiating electrostatic capacitors. The commonly used electrostatic capacitors consist of two metallic plates separating an air gap or dielectric material. As opposed to the battery, as an active component of a circuit, a capacitor is known as a passive component in any ordinary circuit, similar to the resistor and inductor as components of the same circuit. Kaiser (Kaiser 2006) described the capacitor by considering an analogy of a water tank: the capacitor is the tank, any battery is a pump, the resistor and the switch are the valves, and if an inductor is added to the circuit, it is like the moving water in the pipe when connected in a passive circuit with a battery, as depicted in Figure 1.1(a). Figure 1.1(b) shows the induced charges in the dielectric due to the presence of electrical charges on the capacitor's metallic electrodes.

The charging and discharging of a capacitor in a circuit, as shown in Figure 1.1(a), is described by the following differential equations:

$$\frac{dq}{dt} = I = \frac{V_0}{R} e^{\frac{-t}{RC}} \tag{1.1}$$

$$\frac{q}{C} + \frac{dq}{dt} R = 0 \tag{1.2}$$

Where q is the charge stored in time t, V is the potential between the plates, R is the resistance, C is capacitance, I is the current and V_0 is the initial potential between the plates. At the beginning of charging, the total stored charge between the plates is zero. Upon application, a potential charge starts accumulating on the plates. The stored energy in a capacitor is given by the equation:

$$E = \frac{q^2}{2C} = \frac{1}{2} C (V_0)^2 \tag{1.3}$$

The capacitors can be arranged in both series and parallel configurations in different systems, according to the requirements that change the resulting capacitance of the circuit, as shown in Figure 1.2.

DOI: 10.1201/9781003174554-1

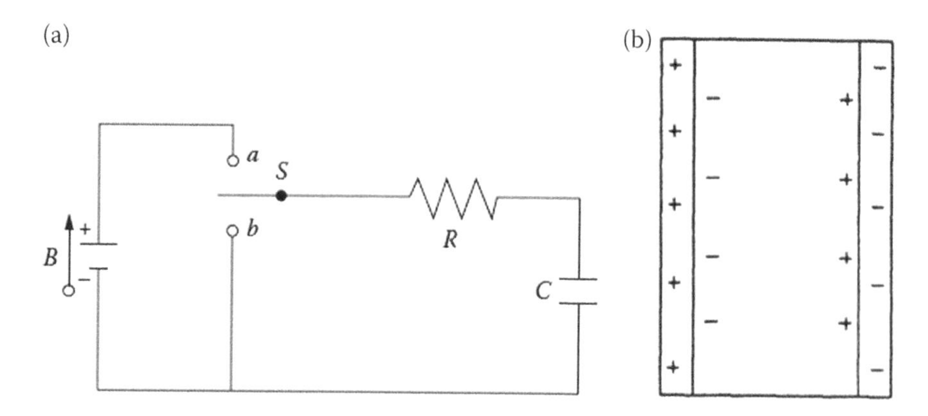

FIGURE 1.1 (a) Electrical circuit in connection with the capacitor, resistor, and a battery. (b) Induced charges on a dielectric sandwiched between metallic plates of a capacitor.

Source: Reprinted with permission from (A. Yu, Chabot, and Zhang 2013), Copyright © 2013, Taylor and Francis.

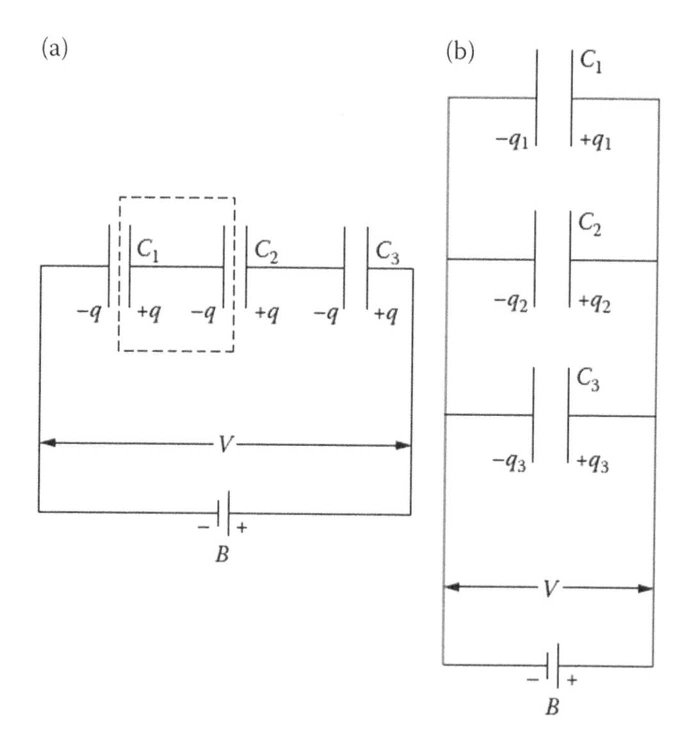

FIGURE 1.2 (a) Series and (b) parallel combination of capacitors.

Source: Reprinted with permission from (A. Yu, Chabot, and Zhang 2013), Copyright © 2013, Taylor and Francis.

The capacitance of a parallel and series combination of capacitors is given below:

$$C_{Eq,parallel} = \frac{Q}{V^o} = \sum_{i=1}^{n} C_i \tag{1.4}$$

and,

$$V_0 = \sum_{i=1}^{n} V_i = Q\left(\sum_{i=1}^{n} 1/C_i\right) = \frac{Q}{C_{Eq,series}} \tag{1.5}$$

Where Q is the total charge. The capacitors connected in series have a total capacitance smaller than any individual capacitor.

1.1.1.2 Applications of Electrostatic Capacitors

It is important to discuss the application of capacitors, where the energy storage becomes crucial. Capacitors are mostly used for energy storage to charge and discharge energy over longer or shorter periods, respectively. Various factors directly influence the working of a capacitor. The heating conditions play a significant role, and most importantly, internal heating raises the temperature of the dielectric and lowers the power factor. An increase in power factor is directly related to the thermal instability at high temperatures. However, the lower temperature may not pose such difficulty. The increase in ambient temperature also decreases the life of a capacitor. The changes in operating temperature also affect the terminal seals made of elastic materials, and a lower temperature may cause internal deformation. Therefore, for a circuit designer, following capacitance specifications are essential for satisfactory operation, as described by the author Kaiser in his book (Kaiser 2006):

1. Tolerances according to specification
2. Capacitance-temperature characteristics
3. Capacitance-voltage characteristics
4. Retrace characteristics
5. Capacitance-frequency characteristics
6. Dielectric absorption
7. Capacitance as a function of pressure, vibration, and shock
8. Capacitor aging in the circuit and during storage

The capacitors are categorized based on the design and the materials used for the specific applications. Below is the mention of different capacitors used for various purposes (described within parenthesis).

1. Ceramic capacitor (for large capacitance and high resistance for insulation)
2. Plastic film capacitor (for high resistance for insulation, lowering of dielectric absorption, lower loss factor over a wide temperature range)

3. Aluminium electrolytic capacitor (filter, coupling, and bypass applications; in the next section, we will detail the electrolytic capacitor)
4. Tantulum capacitor (high-voltage applications)
5. Glass capacitor (with highest performance and reliability features, widely used)
6. Mica capacitor (circuits for precise frequency filtering, bypassing, and coupling)

1.1.1.3 Electrolytic Capacitor

A brief mention about the electrolytic capacitor is important to understand the difference between the electrostatic capacitor and the electrochemical capacitor (supercapacitor). Electrolytic capacitors store energy by an electrostatic charge separation; thus, they are called polarized capacitors. In aluminum electrolytic capacitors, a very thin dielectric layer formed by the anodic oxidation of metals in contact with an electrolyte and the electrolyte forms the cathode. It is separated from the anode by the barrier layer of oxide formed on the anode surface. Solid or non-solid electrolyte acts as counter electrode. However, in the supercapacitor, ionic conductive electrolyte connects the two electrodes for ionic movements. Electrolytic capacitors provide a much higher capacitance-voltage (CV) per unit volume than ceramic capacitors due to a very thin dielectric layer on the anode surface. Aluminum electrolytic capacitors, tantalum electrolytic capacitors, and niobium electrolytic capacitors are the three widely used electrolytic capacitor, as shown in Figure 1.3. For more details, check the relevant references.

The basic differences in working mechanism are depicted in Figure 1.4, where the electrolytic capacitor bridges the other two capacitive behaviors, namely, electrostatic and electrochemical capacitors. The details of the supercapacitor will be discussed in later sections.

1.1.2 Historical Overview of Supercapacitor

After 1st and 2nd generations of electrostatic and electrolytic capacitors, a rapid progress in the development of novel materials led to the development of 3rd generation capacitor to store the charges in a hybrid mode. Helmholtz was the first person to reveal the properties of double-layer capacitance in 1879 (Helmholtz 1879). However, it took many decades before double-layer capacitance was realized as an alternate energy storage. The first supercapacitor with a significantly high capacitance was invented in 1957 by Becker of General Electric (Becker 1957a). It was intended for low-voltage operation to serve as an electrolytic capacitor. The first practical supercapacitor was developed in 1970 with carbon material after further modifications by Boos, a chemist in SOHIO (Boos 1968). Thereafter, European and American countries including Japan, United States, Switzerland, Russia, France, and South Korea started research and commercialization on supercapacitors. A vast global market is mostly occupied by companies like Maxwell of the United States, Japan's NEC, Panasonic, Tokin, and Russian Econd. In 1980s, the industrialization of supercapacitors began by several companies like Generation-1980 NEC/Tokin, Mitsubishi Products, and 1987 Panasonic (Samantara

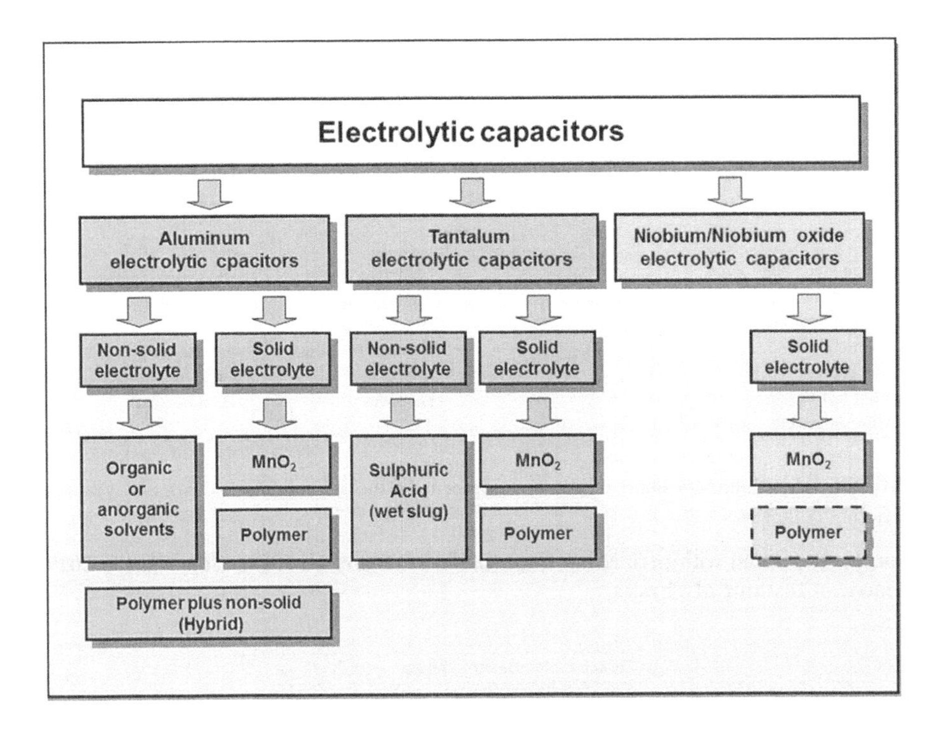

FIGURE 1.3 A wide variety of electrolytic capacitors for various anode metals and the electrolytes.

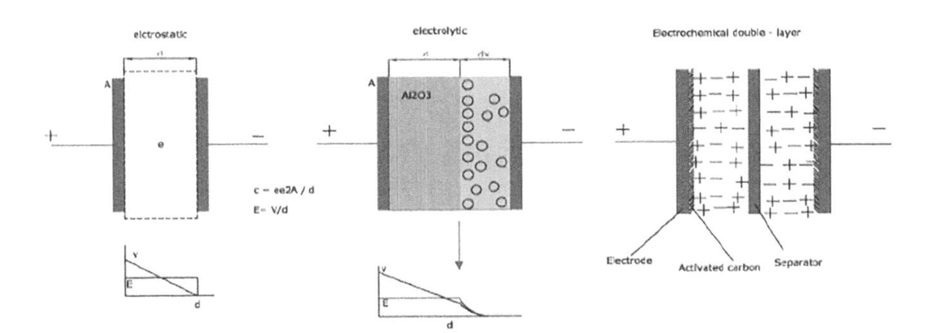

FIGURE 1.4 Schematic diagram showing the mechanisms in electrostatic capacitor, electrolytic capacitor, and electrical double-layer capacitor (supercapacitor) (Jayalakshmi and Balasubramanian 2008a, Int. J. Electrochem. Sci., 3 (2008) 1196–1217).

and Ratha 2018b). In the 1990s, electrochemical capacitors for high power were developed by Econd and ELIT. Many companies, such as NEC, Panasonic, Maxwell, and NESS, are actively involved in the research of supercapacitors, as summarized in Figure 1.5 (Shifei Huang et al. 2019).

Since the supercapacitors were introduced to the global market, the compound annual growth of 39% is noted at the global scale, as depicted in Figure 1.6 (Shifei Huang

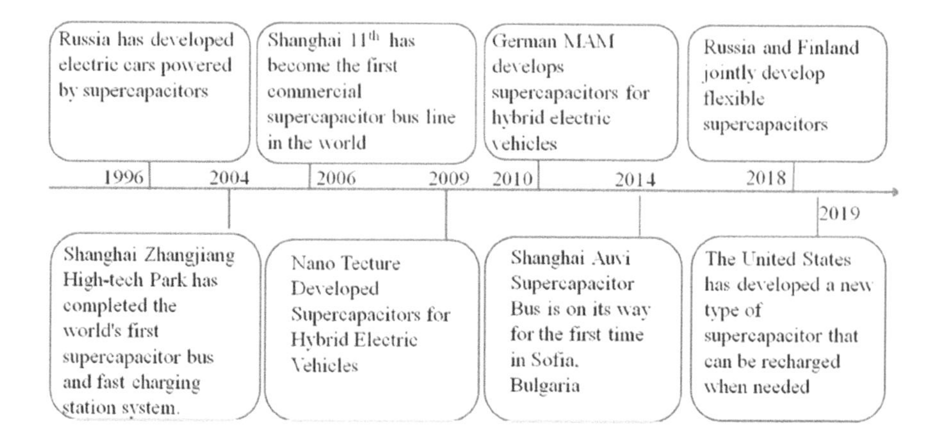

FIGURE 1.5 Summary chart of the development of the supercapacitor over the years in different countries.

Source: Reprinted with permission from (Shifei Huang et al. 2019), Copyright © 2019, American Institute of Physics.

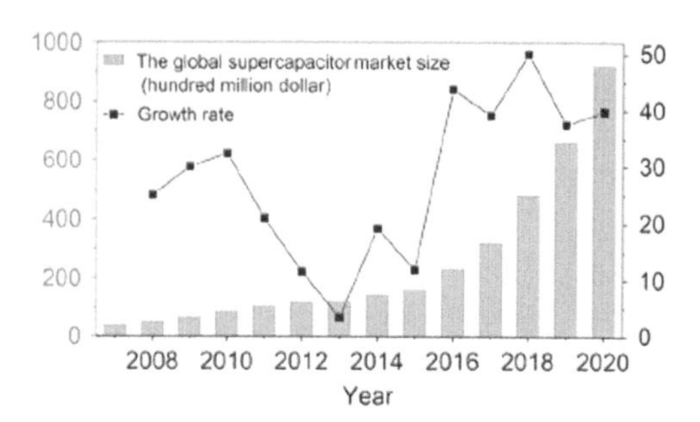

FIGURE 1.6 Global supercapacitor market size with annual growth.

Source: Reprinted with permission from (Shifei Huang et al. 2019), Copyright © 2019, American Institute of Physics.

et al. 2019). With the rapid progress in the development of materials and technological advancements, the performance of supercapacitors is significantly enhanced in recent years. The worldwide rapid development has resulted in a new generation of supercapacitors to be industrialized on a large scale. The demand for the supercapacitors has also been expanded globally on different scales in various industries.

1.2 BASIC PRINCIPLE OF SUPERCAPACITORS AND ENERGY-STORAGE MECHANISM

1.2.1 Basic Principle of Supercapacitors

As in electrostatic and electrolytic capacitors, supercapacitors also use a dielectric material to separate two electrodes. In supercapacitors, charges are stored

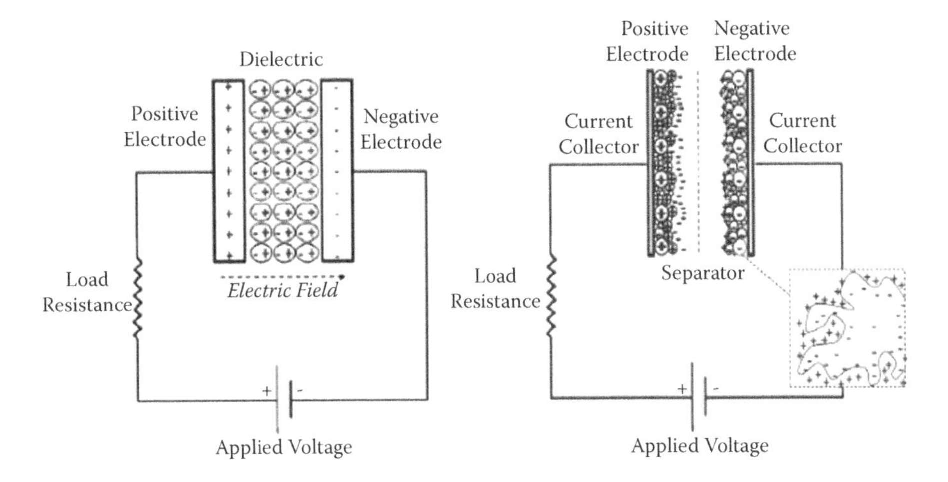

FIGURE 1.7 Schematic diagram to illustrate the charge-separation mechanism in electrostatic capacitor and supercapacitor.

Source: Reprinted with permission from (Prasad et al. 2019), Copyright © 2019, IOP Science.

electrostatically. When a voltage is applied across the terminals, an electric field polarizes the electrolyte, which induces ions to diffuse in the electrodes (Li Zhang and Zhao 2009a). Thus, a resulting electric double-layer is formed at each electrode. The energy-storage capacity depends on the active material and the surface area of the electrodes. Figure 1.7 depicts the fundamental charge-separation mechanism in electrostatic capacitor and supercapacitor.

Several efforts have been made to understand the origin of double-layer structure of capacitance. In this process, models are proposed, as briefly described in the next section.

1.2.1.1 Description of Electrical Double-Layer

Consider that a cell is designated by an electrical circuit consisting of a resister R_s, as a solution resistance and a capacitor, C_d, for double-layer capacitance at the electrode/electrolyte interface in a cell. Figure 1.8 shows Hg/K^+, CF/SCE (standard calomel electrode), cell for an equivalent electrical circuit shown in right side of the figure. The capacitance of SCE, C_{SCE} can be neglected as $C_{SCE} \gg C_d$.

The results for potential steps are similar to the RC circuit, as discussed in the earlier section (equivalent circuit is shown in Figure 1.8). A linearly increasing potential with time at a sweep rate v (in V/s) (see Figure 1.9) is given by,

$$E = vt \tag{1.6}$$

If a triangular wave is applied, then the current changes from vC_d to — vC_d during the forward (increasing E) and the reverse (decreasing E) scan. The plot of such current as a function of potential while C_d remains constant is shown in Figure 1.9.

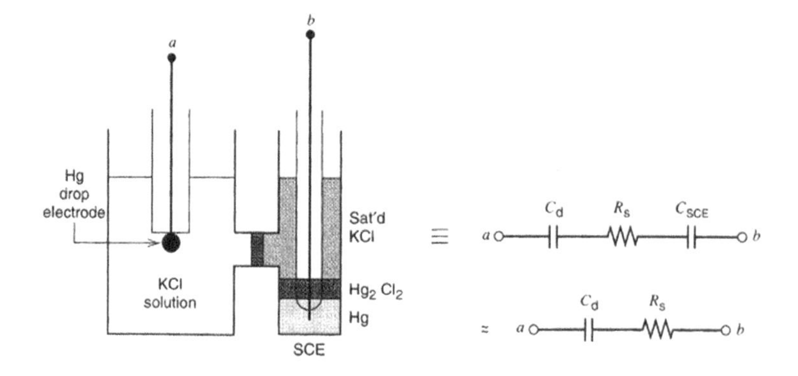

FIGURE 1.8 Left: Two-electrode cell. Right: Linear circuit elements to represent the cell.

Source: Reprinted with permission from (Bard and Faulkner 2001), copyright John Wiley and Sons 2001.

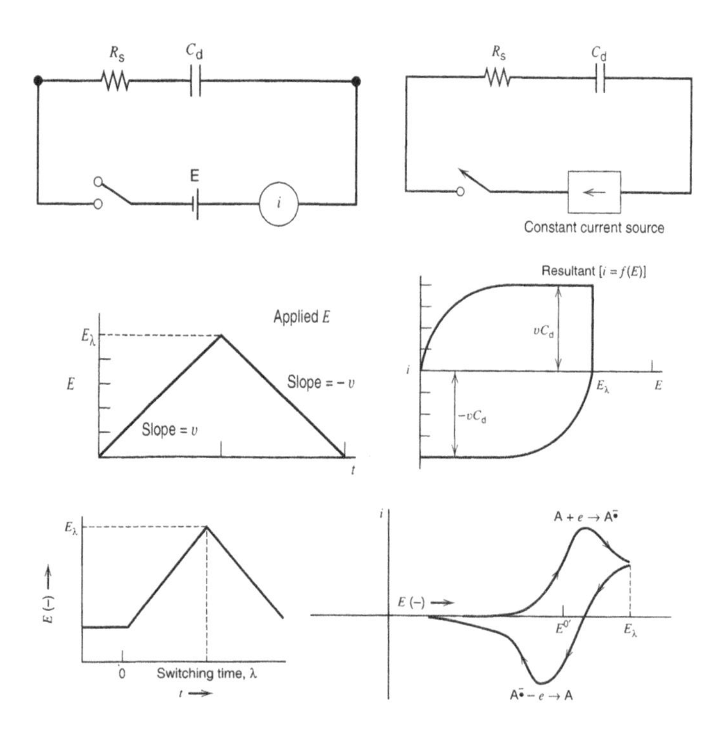

FIGURE 1.9 Circuit diagram for voltage and current sweeps (upper left and right figures). Schematic for different sweeps and corresponding CV curves, potential-time and current-potential plots from a forward and backward linear potential sweep (or a triangular wave) when applied to an RC circuit (lower two, left and right figures).

Source: Reprinted with permission from (Bard and Faulkner 2001), Copyright © 2001, John Wiley and Sons.

Cyclic voltammetry (CV) is a reversal technique used for primary electrochemical studies conducted on any new system. For the Faradic reactions in a cell, if we reverse the potential scan, the current has a shape and peaks similar to that of the forward response/peaks.

1.2.1.2 Peak Current and Charging Current

In a Nernstian (reversible) system, when a linear potential is applied at v (V/s), then potential as a function of time is given by

$$E(t) = E_i - vt \tag{1.7}$$

and the current is given by

$$i = nFAC_O^*(\pi D_O \sigma)^{1/2} \chi^{(\sigma t)} \tag{1.8}$$

The function $\pi^{1/2}\chi^{(\sigma t)}$, hence the current reaches to a maximum, thus, a peak current is given in Figure 1.10.

$$i_P = (2.69 \times 10^5) n^{3/2} A D_O^{1/2} C_O^* v^{1/2} \tag{1.9}$$

Readers are suggested to refer the reference given in Fig. 1.10 for more details of terms given in eqs. 1.8–1.9. Thus, E_p is independent of scan rate and i_p is proportional to $v^{1/2}$ for a reversible wave. Since in a potential sweep, the potential is changing with the time, a charging current flows,

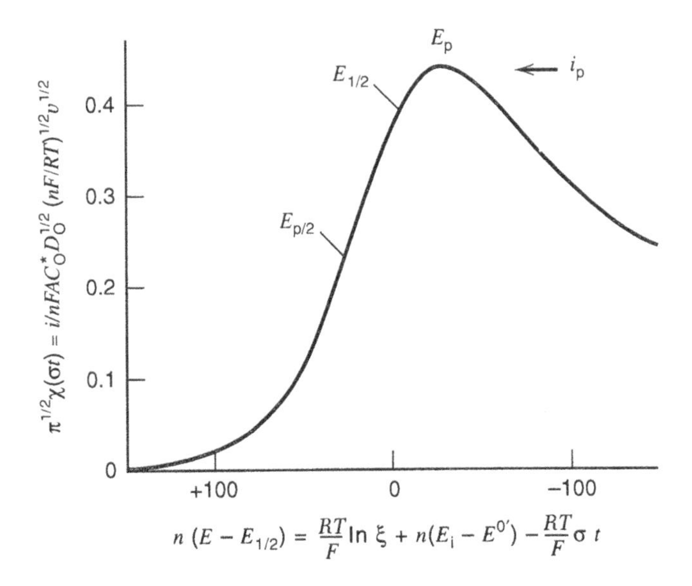

FIGURE 1.10 Potential voltammogram in terms of current function.

Source: Reprinted with permission from (Bard and Faulkner 2001), Copyright © 2001, John Wiley and Sons, Inc.

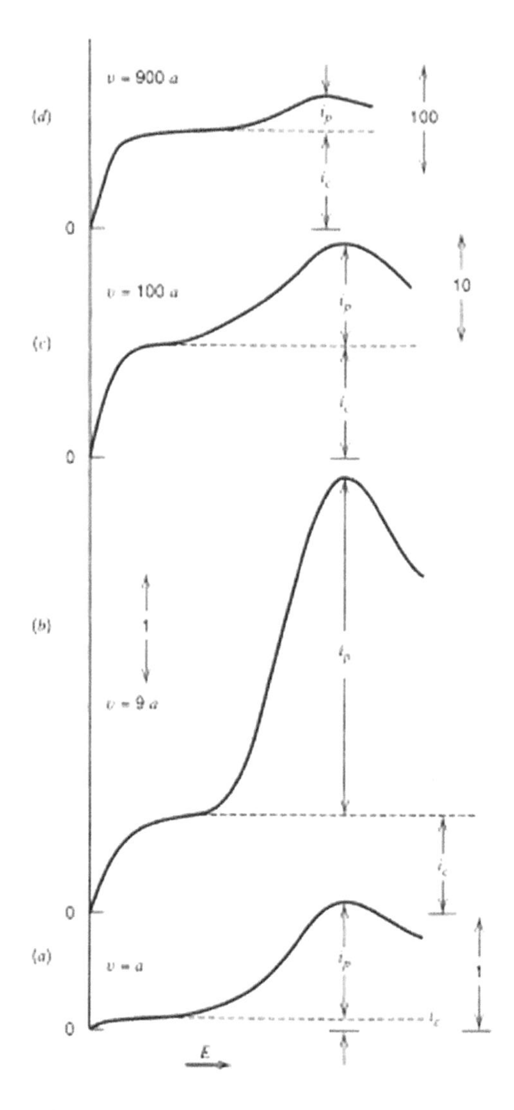

FIGURE 1.11 Plots of double-layer charging at different sweep rates while assuming that C_d is independent of E.

Source: Reprinted with permission from (Bard and Faulkner 2001), copyright John Wiley and Sons 2001.

$$|i_c| = AC_d \upsilon \tag{1.10}$$

and the Faradaic current is measured from the base of a charging current. While i_p varies with $v^{1/2}$ and i_c varies with v, i_c becomes dominant at faster scan rates. It is evident in Figure 1.11 that the surface effects dominate at higher scan rates; it is of least importance at a lower scan rate, where C_d is independent of potential.

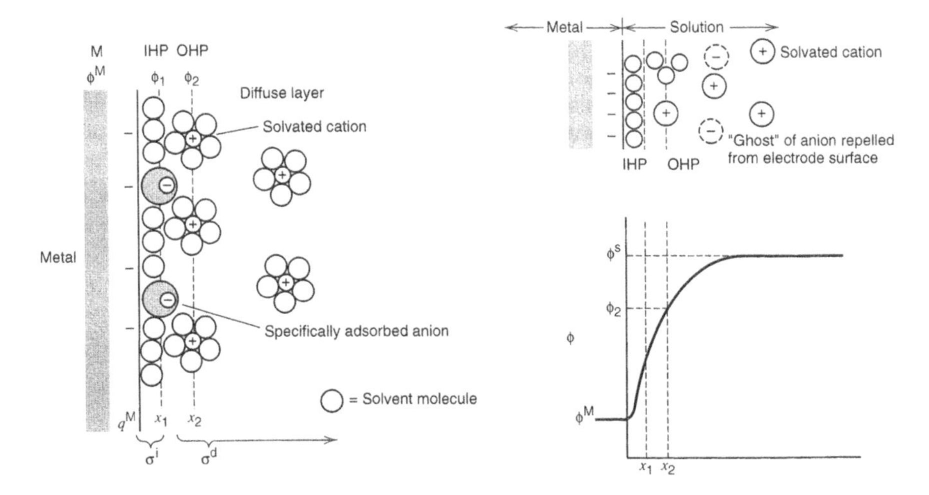

FIGURE 1.12 Schematic illustration of double-layer formation at the surface of metal. Right figure shows the potential profile with the distance from the electrode.

Source: Reprinted with permission from (Bard and Faulkner 2001), copyright © 2001, John Wiley and Sons, Inc.

The variation in magnitudes of both the charging current, i_c and the Faradaic peak current, i_p, is shown in Figure 1.11.

Figure 1.12 further explains the formation of a double-charge layer. "Several layers" are considered near the solution side of the double-layer. The inner layer closest to the electrode surface has specifically adsorbed ions and solvent molecules (Figure 1.12); this layer is named the compact, Helmholtz, or Stern layer. The potential profile of double-layer near the electrode region with the thickness (distance) of the diffuse layer across is shown in the right image. The ion concentration in the solution decides the thickness of the diffuse layer.

The inner Helmholtz plane (IHP) is the distance x_1 of the specifically adsorbed ions with the total charge density in the inner layer as $\sigma^i\,(\mu C/cm^2)$. The outer Helmholtz plane (OHP) is a distance x_2 from the metal having the nearest solvated ions only. Since these ions are nonspecifically adsorbed, they can distribute in a three-dimensional region on any thermal-agitation in the solution, which is a diffuse layer. The formed diffuse layer can extend from the OHP into the solution. If the diffuse layer has an excess-charge density σ^d, then the net excess charge density of the double layer on the solution side, σ^S, is given by

$$\sigma^S = \sigma^i + \sigma^d \tag{1.11}$$

The structure of the double-layer can significantly affect the rate of processes. Moreover, as mentioned earlier, the thickness of the diffuse layer depends on the solution's ion concentration. Therefore, for the electrode reactions involving very low concentration of the electroactive species, the Faradaic current for any reduction or oxidation reaction will be lesser than the charging current. In the absence of electroactive substance, a double-layer charging current flows with a current generated due to the oxidation-reduction of stray electroactive substances, such as (a) the electrode, (b) impurities from heavy metals, and (c) the solvent or electrolyte. When the Faradaic component of the current is small, the non-Faradaic current, which is the charging or capacitive current, dominates the net current. The charge contained on the double-layer is given by,

$$q = -C_i A(E - E_Z) \tag{1.12}$$

where C_i is the capacitance of the double layer on an electrode area of A. $E - E_z$ is the difference in potential of the electrode and potential where the excess charge on the electrode is zero (potential of zero charge, PZC). The charging current will be,

$$i_c = \frac{dq}{dt} = C_i(E_Z - E)\frac{dA}{dt} \tag{1.13}$$

Further, models are proposed to evaluate exact qualitative relation. The next section describes the successive models.

1.2.1.3 The Helmholtz Model

About the charge separation at electrode/electrolyte interfaces, Helmholtz (Helmholtz 1879) was the first to propose that the counter-charge in a solution forms two layers of charge of opposite polarity having a separation of molecular order that resides at the electrode's surface. Therefore, this structure can also be described as equivalent to the configuration of parallel-plate capacitor given by:

$$\sigma = \frac{\varepsilon\varepsilon_0}{d}V \tag{1.14}$$

where, charge density, σ, and the voltage drop, V, at the plates and ε is the dielectric constant of the medium, ε_0 is the permittivity of free space and d is the spacing between the plates. The differential capacitance is,

$$\frac{\partial\sigma}{\partial V} = C_d = \frac{\varepsilon\varepsilon_0}{d} \tag{1.15}$$

which apparently is constant as opposed to the observed behavior in solution interfaces, e.g., where a nonlinear variation of differential capacitance can be observed with the potential, as depicted in Figure 1.13.

Thus, the proposed model does not support fully the experimental observations.

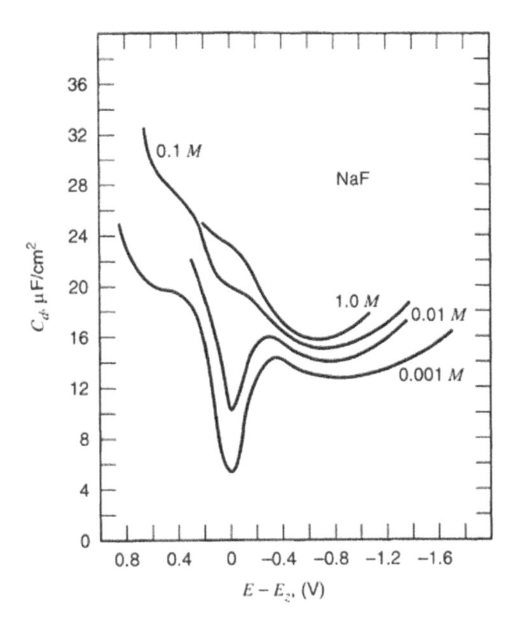

FIGURE 1.13 Variation of differential capacitance with respect to potential for various concentration of NaF concentration.

Source: Reprinted with permission from (Bard and Faulkner 2001), copyright © 2001, John Wiley and Sons Inc.

1.2.1.4 The Gouy-Chapman Theory

Further, the theory proposed by Gouy-Chapman explained that an interplay occurs between the charge transfer at the metallic electrode, depending on polarity and involved thermal processes to randomize them, thus leading to the formation of a finite charge thickness. This model therefore considered the involvement of a diffuse layer of charge in the solution. The idea of the diffuse layer was proposed by both Gouy (Gouy 1910) and Chapman (Chapman 1910) independently.

The potential profile in the diffuse layer at different potential is given in Figure 1.14(a), showing the potential decay away from the surface. The predicted V-shaped capacitance function has a resemblance that was observed in NaF at low concentrations, shown in Figure 1.14(b). However, the model does not explain a constant capacitance at large potentials and at high electrolyte concentrations where the valley or dip at the PZC disappears. The actual capacitance value is measured much lower than the value predicted from the proposed model.

1.2.1.5 Stern's Modification

In the Gouy-Chapman model, an unlimited rise in differential capacitance (Figure 1.14) is due to the unrestricted ions in the solution, which can approach the surface in an arbitrary manner. Thus, the separation between the charges in

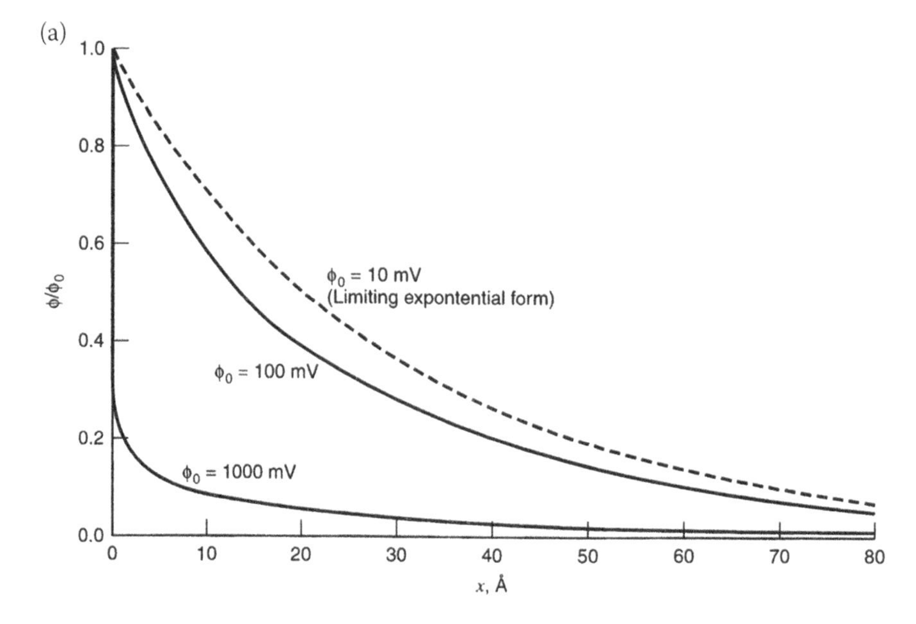

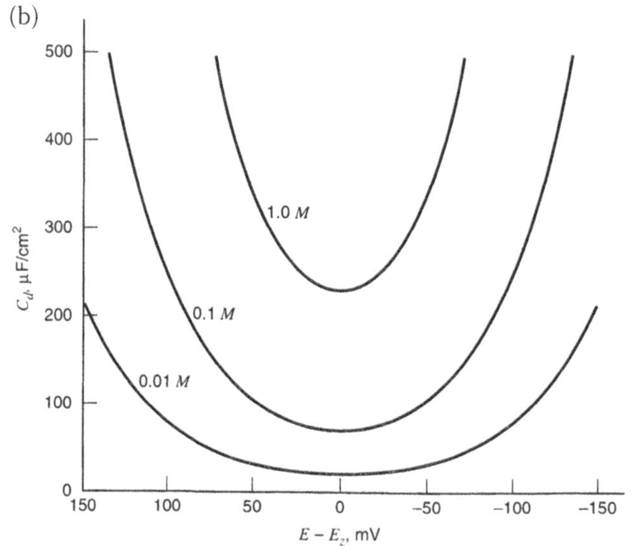

FIGURE 1.14 (a) Potential vs. diffuse layer thickness and (b) capacitance vs. voltage curve.

Source: Reprinted with permission from (Bard and Faulkner 2001), copyright © 2001, John Wiley and Sons Inc.

metallic and solution-phase zones reduces toward zero at high polarization condition or high electrolyte concentration. This view cannot be realistic under this condition, when it cannot approach the metallic surface closer than its ionic radius due to its finite size. Moreover, with more concentrated electrolytes (or at larger polarizations), the charge in solution gets crowded near the boundary at x_2 as in the Helmholtz model. On the other hand, with low electrolyte concentration, the diffuse layer (thickness) becomes larger compared to x_2, so prediction for the capacitance for potentials near the PZC remains the same. The plane at x_2 is called the outer Helmholtz plane (OHP), and this model was first suggested by Stern (Stern 1924).

The capacitance is made up of two series components, as defined below.

$$\frac{1}{C_d} = \frac{1}{C_H} + \frac{1}{C_D} \tag{1.16}$$

If C_H is the capacitance of the charges at the OHP, whereas C_D is the capacitance due to the diffuse layer charge. As mentioned earlier, the C_H does not depend on the potential, and C_D varies in a valley shaped near the PZC in a low electrolyte concentration system. At larger electrolyte concentrations or high polarizability, C_D becomes large for a negligible C_d and only a constant capacitance of C_d appears for all ranges of potential, including PZC. Figure 1.15 is a schematic to represent this behavior.

1.2.1.6 Pseudocapacitance

The concept of pseudocapacitance is developed to increase the capacitance of an electrochemical supercapacitor, where electrodes are developed from electrochemically active materials and explored to provide an additional capacitance along

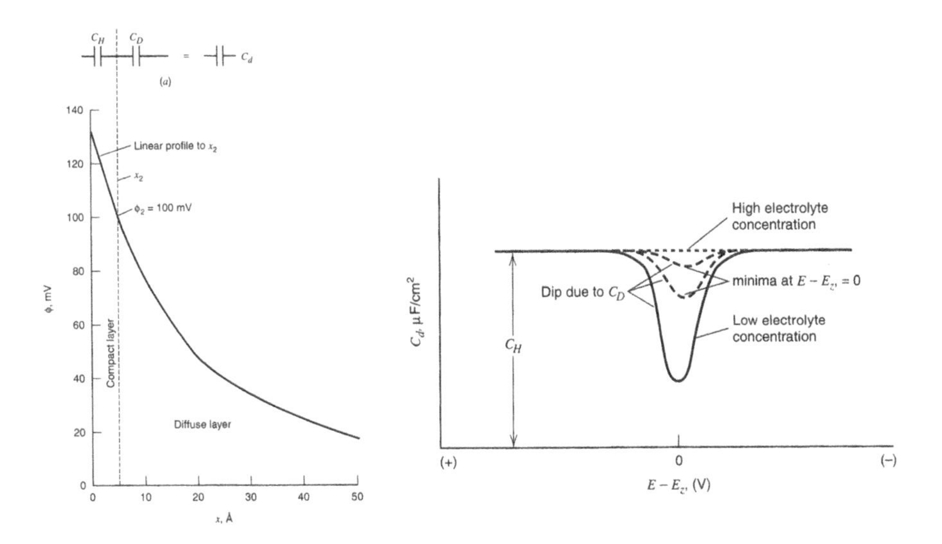

FIGURE 1.15 Potential distribution and diffuse capacitance variation with respect to distance from electrode and potential, respectively.

Source: Reprinted with permission from (Bard and Faulkner 2001), copyright © 2001, John Wiley and Sons Inc.

with the double-layer capacitance. The charge-storage mechanism in pseudocapacitive process is different from double-layer capacitance in which a surface-based Faradic charge transfer occurs via electrochemical reduction–oxidation reactions that are thermodynamically and kinetically favored. So far, metal oxide and conducting polymers that exhibit protonation reactions and absorption into the polymer matrix are widely used. High surface area carbon as electrodes are best used to increase the pseudocapacitive component. Asymmetric configurations in electrochemical supercapacitors are preferred, where separate materials for the anode and cathode exhibit both the higher capacitance and voltage stability. The total capacitance (C_T) of the electrode can be the combined double-layer capacitance (C_{dl}) and pseudocapacitance (C_{pc}) (A. Yu, Chabot, and Zhang 2013):

$$C_T = C_{dl} + C_{pc}(E) \tag{1.17}$$

$$= C_{dl} + \frac{n^2F^2}{RT} dC^o_{Ox} \tag{1.18}$$

$$= \frac{exp\left(\frac{nF}{RT}(E^o_{Ox/Rd} - E) + g\frac{C_{Ox}}{C^o_{Ox}}\right)}{\left[1 + exp\left(\frac{nF}{RT}(E^o_{Ox/Rd} - E) + g\frac{C_{Ox}}{C^o_{Ox}}\right)\right]^2 + g\ exp\left(\frac{nF}{RT}(E^o_{Ox/Rd} - E) + g\frac{C_{Ox}}{C^o_{Ox}}\right)} \tag{1.19}$$

Readers are suggested to refer the reference given in Fig. 1.16 for more details of terms given in eq. 1.19. Figure 1.16 shows the plot of total capacitance with the reversibly applied electrode potential in cyclic voltammetry.

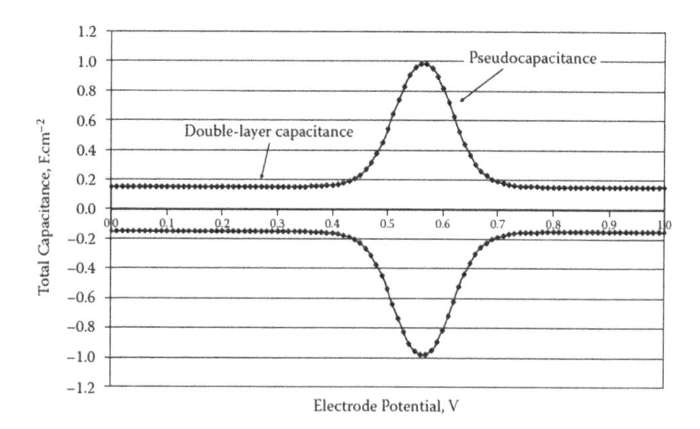

FIGURE 1.16 The total capacitance as described in Equation 1.19 as a function of electrode potential.

Source: Reprinted with permission from (A. Yu, Chabot, and Zhang 2013), Copyright © 2013, Taylor and Francis.

Pseudocapacitance can be produced from various types of redox reactions, such as ions intercalation, underpotential deposition, and electrically conducting polymers, bulk redox reactions, etc., as given below.

1. Pseudocapacitance induced by underpotential deposition
2. Pseudocapacitance induced by lithium intercalation
3. Pseudocapacitance induced by redox couples
   a. Pseudocapacitance by dissolved redox couples
   b. Pseudocapacitance by undissolved redox couples
4. Pseudocapacitance induced in electrically conducting polymer (ECP)
5. Coupling of differential double-layer and pseudocapacitance

1.2.2 Energy-Storage Mechanism

The supercapacitor operation is associated with the energy distribution and storage of the electrolytic ions in contact with the electrodes. Based on the method of charge storage, supercapacitors have been divided into three classes: pseudocapacitors, electrochemical double-layer capacitors (EDLCs), and hybrid supercapacitors, as shown in Figure 1.17. EDLCs are composed of an electrolyte,

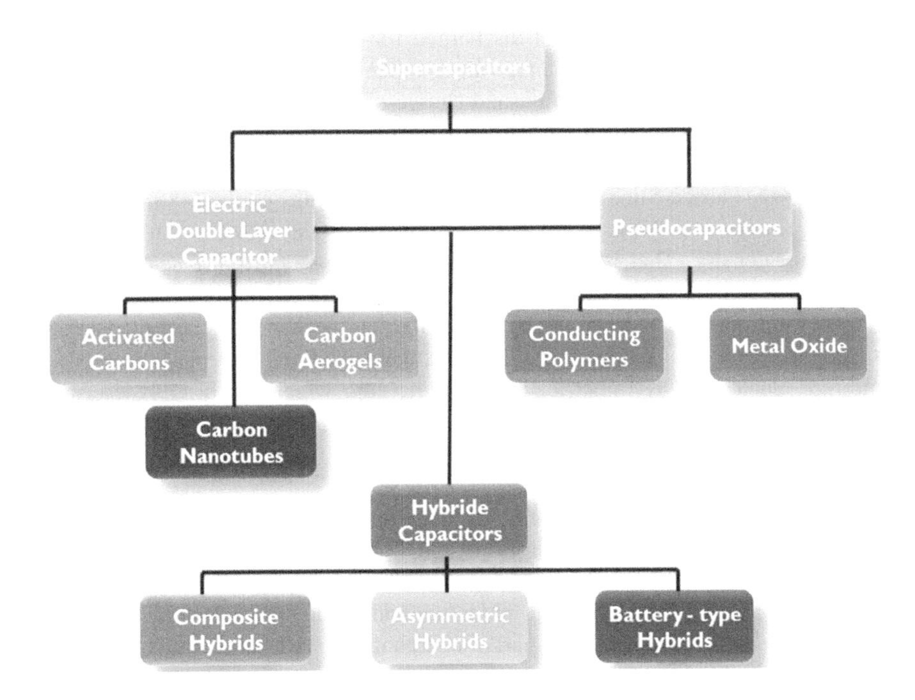

FIGURE 1.17 Classification of the supercapacitors according to the charge storage mechanism.

Source: Reprinted with permission from (Jadhav, Mane, and Shinde 2020), Copyright © 2020, Springer, Cham.

a separator, and two high surface-area carbon-based electrodes. EDLCs also termed as non-Faradic charge-storage process for storing charge electrostatically, which refrain from any transfer of charge between electrolyte and the electrode (Kiamahalleh et al. 2012), (Jayalakshmi and Balasubramanian 2008b). An electrochemical double-layer is responsible for the charge storage in the case of the EDLC supercapacitor. On application of voltage, charge accumulates on the electrode surfaces due to the attraction of opposite charges upon variation in potential, resulting in diffusion of the electrolytic ions on the porous surface of the opposite charged electrode. At the electrode-electrolyte interface, a double-layer forms to limit the recombination of the ions. The electrical double-layer, in combination with the increase in the specific surface area and decreased gap between the electrode, allows EDLCs to reach a much larger energy density as compared to the conventional capacitors (Kiamahalleh et al. 2012), (H. Choi and Yoon 2015). Furthermore, as a result of involvement of only surface of the electrode material for charge storage, the charge uptake and release is very rapid resulting in high power density. Due to the absence of any Faradic process for charge storage, no chemical reaction occurs thus, resulting in the elimination of the swelling observed in the electrode active material. This phenomenon is very common in the case of batteries due to the presence of Faradaic reaction involving the electrode active material for the charge storage. Some of the noticeable differences between EDLCs and batteries are (i) EDLCs can be charged and discharged for millions of cycles without any significant decrease in the capacitance value, unlike batteries whose charge-storage capacity is degraded with a few thousand charge-discharge cycles. (ii) The charge-storage mechanism is nowhere related to the solvent of the electrolyte for EDLCs; however, for Li-ion batteries, the solid electrolyte inter phase is contributed by the electrolyte (Patrice Simon and Gogotsi 2008), (Hong Li et al. 2009). However, attributed to the non-Faradaic surface-charging process, EDLCs can deliver a little energy density. That is the primary reason why the current research on supercapacitors is mainly based on improving the charge-storage capacity (energy density of the supercapacitor). Based on different mechanisms, supercapacitors are categorized as pseudocapacitors, hybrid supercapacitors, supercapacitors with composite electrodes, asymmetric, battery-type, etc., which are briefly described in the next sections.

1.2.2.1 Pseudocapacitors

Unlike EDLCs, which store charge electrostatically through a non-Faradaic process, charge storage in the pseudocapacitors involves Faradic interaction (charge transportation) between the electrolyte and the electrode (Mohapatra, Acharya, and Roy 2012). On application of a potential, reduction and oxidation occurs involving the electrode material, resulting in transport of charge through the double-layer, causing the emergence of a Faradic current flowing in the supercapacitor. In case of the pseudocapacitors, since the charge storage involves redox interaction, the energy density of the Faradic process in pseudocapacitors allows them to achieve larger energy densities and specific capacitance as compared to EDLCs. Examples of materials storing charge through this mechanism include metal oxides, conducting polymers, metal organic frameworks, polyoxometalates, MXenes, metal nitrides, metal sulfides, etc. However, since these materials undergo reversible redox reaction, they also face limited cyclic stability and low power density as compared to EDLCs (S. M. Chen et al. 2014).

1.2.2.2 Hybrid Supercapacitors

EDLCs exhibit a good cyclic stability and high power performance; on the other hand, pseudocapacitance is associated with high energy density and specific capacitance. In a hybrid system containing materials both of pseudocapacitance origin and EDL origin both advantages can be combined (T. Chen and Dai 2013). An appropriate combination of electrodes can result in an increase of the cell voltage, resulting in an improvement in power and energy densities. Numerous kinds of combinations have been studied so far, involving the negative as well as the positive electrodes in inorganic and aqueous electrolytes. In general, the electrode involving Faradaic reactions contributes to increased energy density with reduced cyclic stability, which is the most significant challenge associated with hybrid devices when compared to EDLCs (Naoi and Simon 2008). Presently, researchers are concerned with three different classes of hybrid supercapacitors distinguished by the configuration of the electrodes: asymmetric, composite, and battery-type.

1.2.2.3 Composite Electrodes

Composite electrodes combine materials of pseudocapacitive origin with carbonic materials within a single electrode system that will be associated with chemical as well as physical charge storage. Carbon-based materials are responsible for the presence of charge storage of capacitive double-layer origin combined with large specific surface area enhancing the interfacial area between the electrolyte and the pseudocapacitive material. The pseudocapacitive material is responsible for increasing the specific capacitance by Faradaic reaction (Kiamahalleh et al. 2012).

1.2.2.4 Asymmetric Hybrids

Asymmetric hybrids syndicate non-redox and redox charge-storage mechanism having both EDLC electrode as well as a pseudocapacitor electrode. In the setup, the carbon material serves as the negative electrode while the material of pseudocapacitive origin acts as the positive electrode.

1.2.2.5 Battery Type

The battery type hybrid is designed by combining two different electrodes, similar to the asymmetric hybrids; the only difference is that, in this case, one electrode is of supercapacitor type and the other is a battery electrode. This configuration helps to use the properties of both batteries and supercapacitors in a single cell (Kiamahalleh et al. 2012),(Iro, Subramani, and Dash 2016).

The battery type, capacitive type, and pseudocapacitive type materials exhibit different reaction mechanisms. As can be seen in Figure 1.18 and 1.19, in supercapacitors termed as EDLCs, charge storage occurs in the oppositely charged double-layer formed at the interface of the electrode and the electrolyte. This type of charge storage occurs only at the electrode material surface; hence, largely porous materials associated with high surface area are essential for reaching

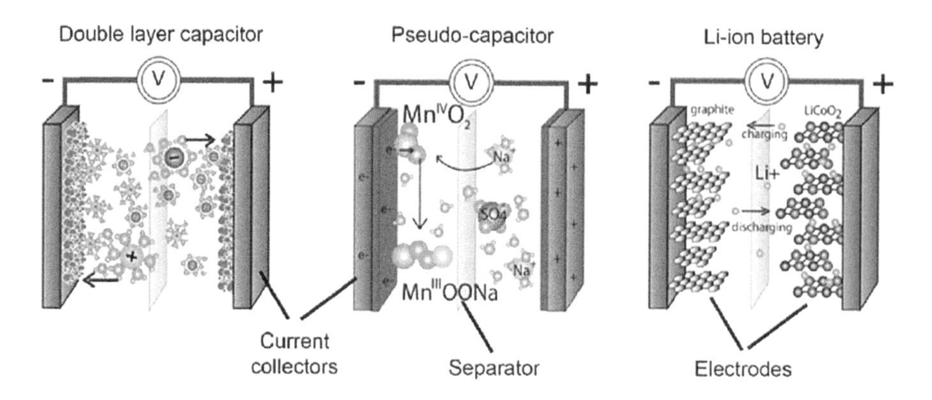

FIGURE 1.18 The basic charge storage difference between the ultracapacitor, pseudocapacitor, and rechargeable batteries.

Source: Reprinted with permission from (Gulzar et al. 2016), copyright © 2016, Royal Society of Chemistry.

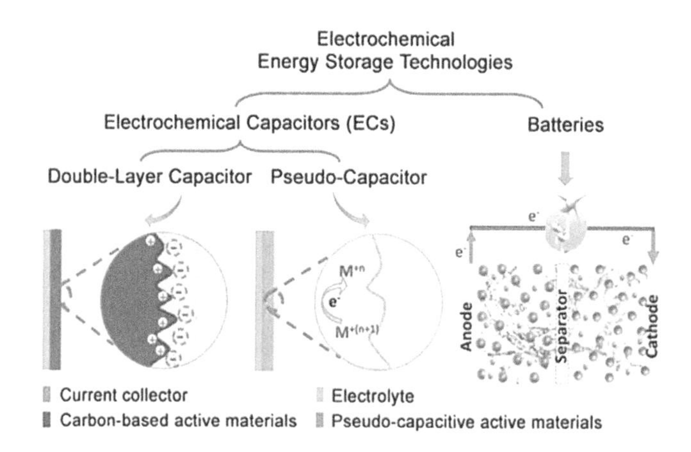

FIGURE 1.19 Different type of pseudocapacitance.

Source: Reprinted with permission from (J. Liu et al. 2018), copyright © 2018, John Wiley & Sons.

significant charge storage. In this case, no Faradaic reaction is involved in the charge-storage process. All the carbon electrode falls under this category. Unlike the EDLCs, pseudocapacitors store charge through highly reversible Faradaic interaction of the electrode active material with the electrolyte. In this case, charge transfer occurs between the electrolyte and the electrode. However, the dependence of charge storage with the voltage is such that the derivative of charge stored dQ/dt shows capacitive behavior. Also, in this case, the Faradaic interaction is not associated with any chemical change of the electrode material. This is the main feature that differentiates pseudocapacitor material from the battery-type material.

As can be seen in Figure 1.18 and 1.19, a chemical reaction is going on between the electrode and the electrolyte as the charge and discharge occurs in case of battery. This reaction is not limited to the surface only but also extends to the bulk of the electrode material. This process is not very fast and is not as reversible, as in case of the EDLCs or pseudocapacitors, leading to the lesser power density. However, as the bulk of the electrode material is involved in storing charge, the amount of charge stored and hence the energy density is much larger for batteries as compared to EDLCs or pseudocapacitors.

1.3 BASIC STRUCTURE AND COMPONENTS OF A SUPERCAPACITOR

The electrode materials in electrochemical supercapacitor are characterized for their electrochemical properties in a three-electrode or two-electrode configuration. Working, counter, and reference electrodes are the components of a three-electrode system, as shown in Figure 1.20 (a). In a three-electrode setup, a reference electrode is used to measure the applied voltage across the working electrode. As described in the earlier section about the equivalent circuit for a cell, the circuit elements are arranged for the electrochemical setup along with a work station.

Figure 1.20 (b) shows a two-electrode electrochemical system that uses a pair of electrodes; electro-active materials are the current collector, and a separator sandwiched between the parallel electrodes is saturated with an electrolyte. The electrically insulating (dielectric) separator is an ion-permeable dielectric porous thin membrane to allow the transfer of electrolytic ions as when the two electrodes remain electrically separated to prevent electrical short circuits. The whole system comprises a cell having two electrodes: i.e., positive and negative electrodes.

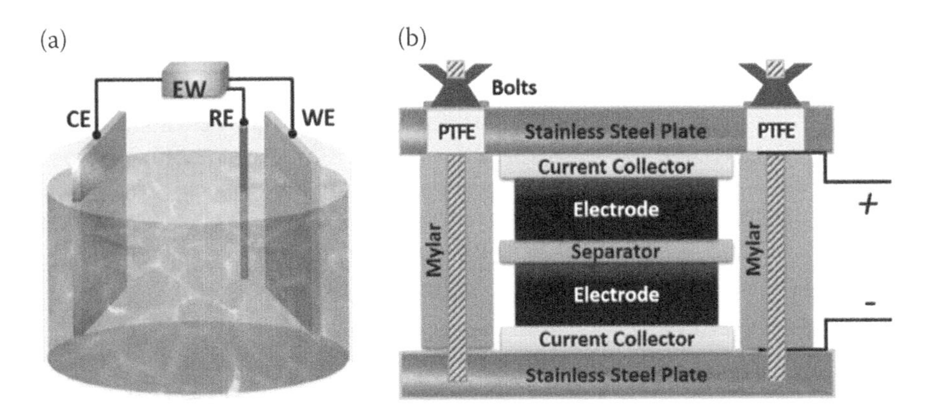

FIGURE 1.20 Schematic showing construction of (a) the three-electrode electrochemcial setup and (b) the two-electrode electrochemcial setup.

Source: Reprinted with permission from (Zhong Wu et al. 2017), Copyright © 2017, Wiley and Sons Inc.

1.4 SYMMETRIC AND ASYMMETRIC SUPERCAPACITORS

Two-electrode configuration can be obtained through asymmetric and symmetric systems on the basis of different types of electrode materials. In symmetric supercapacitors, similar voltage is employed on two electrodes in cells configured symmetrically having electrical double-layer capacitance (EDLC), as well as pseudo-capacitive electrode material. In the asymmetric configuration, the electrodes are associated with one pseudocapacitive electrode and the other EDLC electrode acts as the positive and negative electrodes, respectively. The asymmetric devices garnered a lot of attention as a result of large energy density and a large power density associated with enhanced operating potential. An architecture properly optimized will enable to improve the performance further in an asymmetric supercapacitor.

1.5 PLANAR AND NON-PLANAR SUPERCAPACITORS

Portable electronic devices and microelectromechanical systems have generated a requirement of microsupercapacitors as a microscale energy-storage devices. The microsupercapacitors have demonstrated the feasibility to be directly integrated into various devices. The advantages associated with micron-size devices are smaller diffusion distance for the electrolyte in all the dimensions of the electrodes, thus covering a larger surface area of the electrodes. Secondly, small-scale energy-storage systems have shown higher specific capacitance and higher power than conventional supercapacitors. Planar microsupercapacitors have several advantages; for on-chip integration, the two planar electrodes are in the same plane, with much reduced interplanar spacing to minimize the ion diffusion path length meant for transfer of ions; they can be controlled easily by incorporating various parameter, including kinetics. Planar microsupercapacitors attain a high amount of discharge energy (delivery) at large rates of charge-discharge. The two electrodes in planar configuration are isolated electrically according to their design, in the absence of any separator, as in nonplanar supercapacitors. The interdigitated structure in the plane can also be extended vertically for incorporation of more active materials.

Conventionally used nonplanar capacitors, as described earlier, can have two or three-electrode systems with a large variation in the electrode materials. The nonplanar geometry of supercapacitors does not provide solutions to onchip integration or a way toward development of a miniaturized system.

1.6 ELECTROLYTES

Electrolytes have an important role to play in the performance of an electrochemical capacitor. The system's internal resistance is mainly attributed to the ionic conductivity of the electrolyte and electrolyte–electrode interactions; this conductivity increases as the instability of an electrolyte at different operating temperatures, as well as chemical instability at high rates, and hence reduces the cycle life. The electrolytes should exhibit high electrochemical and chemical stabilities for larger potential windows and high performance characteristics. Higher voltage in electrochemical capacitors can cause a breakdown in electrolyte material, thus limiting potential within a specific range only. A sharp current tail appears at the end of the applied voltage range due to the decomposition of the electrolyte.

1.6.1 Aqueous Electrolytes

In aqueous electrolytes, the sources of ions commonly being used are potassium hydroxide, potassium chloride, and sulfuric acid because of high ionic conductivity. The issues such as corrosion and a low potential window are the common disadvantages of aqueous electrolytes. Water decomposition in aqueous electrolytes results in hydrogen and oxygen evolution at around 0 V and 1.2 V because of the poor electrical stability of water, as shown in Figure 1.21. Therefore, for aqueous systems, the potential window is mostly limited at around 1 V.

1.6.2 Organic Electrolytes

Organic electrolytes are known for their larger potential window in the range of 2.2 to 2.7 V, which directly implies to the larger energy and power densities for most market requirements. Higher resistances in organic electrolyte can be tailored via a good design of optimize capacitance. However, the large internal resistance induces charge leakage across the double-layer interface due to the presence of water.

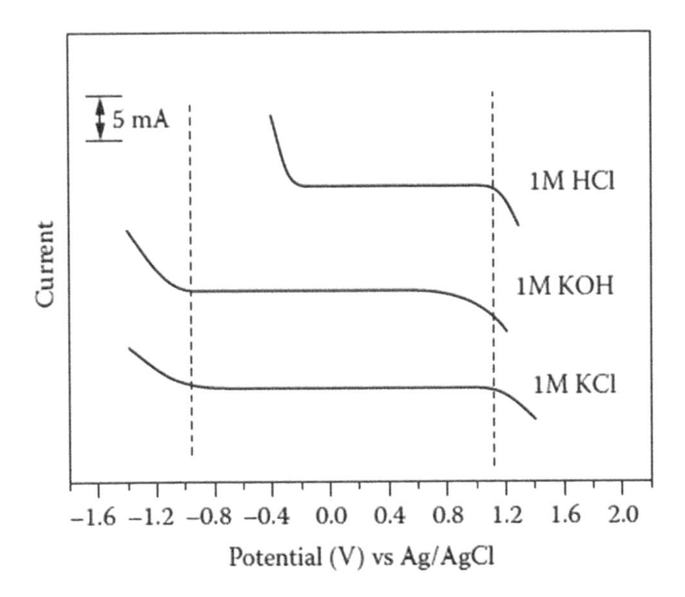

FIGURE 1.21 CV from platinum electrode, showing gas evolution from different aqueous electrolytes during electrochemical testing.

Source: Reprinted with permission from (M. S. Hong, Lee, and Kim 2002), Copyright © 2002, IOP Publishing.

1.6.3 Ionic Liquids

High chemical stability of ionic liquids allows operation at high voltage windows up to 5 V. Low conductivity of ionic liquids is a major drawback at room temperature. Ionic liquids-based electrolytes have high thermal stability for operation in high temperature conditions.

1.6.4 Solid State Polymer Electrolytes

In solid-state polymer electrolytes, both the electrolyte and separator are used as a single component. A solid polymer film is prepared from a gel electrolyte, a liquid electrolyte held through capillary forces into a microporous polymer layer. Gel electrolytes are also used with aqueous, organic, and ionic liquids, depending on the applications.

2 Electrochemical Measurements for Supercapacitors

2.1 MEASUREMENT OF THREE-ELECTRODE SYSTEM

In a three-electrode setup, a reference electrode is kept aloof from the counter and is connected to another electrode termed as the working electrode. The electrode's position is responsible for the measurement of a point in the vicinity of the working electrode, which has both working (W) and working sense (WS) electrical connections, as illustrated in Figure 2.1 (Echendu et al. 2016). Three-electrode configurations only monitor the change in potential of the working electrode as compared to the reference electrode, which is half of the full cell. These affects are not dependent on the variations that take place at the counter electrode. Henceforth, three-electrode electrochemical configuration is the more commonly implemented setup. An alternate configuration associated with the three-electrode configuration invovles monitoring the voltage changes between the reference and working sense and the reference and the counter sense, simultaneously. Here, the reference is linked with the reference electrode, counter sense and counter are linked to the counter electrode, while working and working sense are linked to the working electrode. In a single experiment, the results from both the half cells are obtain through this type of configuration (Echendu et al. 2016).

2.2 MEASUREMENT OF TWO-ELECTRODE SYSTEM

Two-electrode configuration is a widely used setup. It uses a simpler setup but may have complex results and corresponding analysis. The design for two-electrode setup involves sense and current linked together: working sense (WS) and working (W) are linked to an electrode termed as working and the reference (R), as well as the counter (C) electrode, are linked to another electrode (Echendu et al. 2016). Figure 2.1(a) illustrates the setup of a two-electrode cell. In a two-electrode experiment, the entire electrochemical cell is comprised of electrolyte, working electrode, and counter electrode, which means the sense leads monitor the entire voltage drop across the current. Figure 2.1(b) shows a tracing associated with the whole-cell potential for a two-electrode setup to monitor the voltage drop across the entire cell in which the working sense link is at point A and a reference is connected at the point E. In general, two-electrode setups are useful for monitoring a significant voltage drop across the entire cell, including batteries, supercapacitors, fuel cells, and so on. They are also useful in a situation where the drift associated

DOI: 10.1201/9781003174554-2

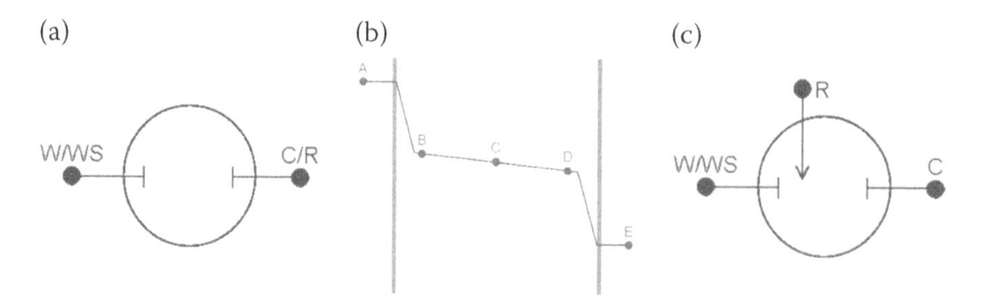

FIGURE 2.1 (a) 2-electrode configuration; (b) A potential map across a whole cell; (c) 3-electrode cell setup.

with the counter-electrode potential is smaller during the experiment; the systems result in very small gains within a short timescale. (Echendu et al. 2016).

A three-electrode setup consists of a counter electrode (CE), reference electrode (RE), and a working electrode (WE) (Figure 2.1(c)). In this configuration, we measure the current between the working and the counter electrode and measure the potential difference between the working and the reference electrode. Hence, in a three-electrode configuration, the current and the potential of only the working electrode can be monitored (Figure 2.2).

However, in a two-electrode configuration, both the electrodes are working electrodes (there is no reference electrode or counter electrode), and the current and potential of the entire system is monitored. It is noteworthy that the three-electrode configuration is best for understanding the characteristics of an electrode material. However, the three-electrode configuration does not resemble the real device architecture and the specific capacitance calculated by employing a three-electrode system, which is four times the value that would be obtained if the same electrode material were cast as a device. In this matter, the two-electrode system resembles the real device more closely; hence, this configuration should be used to obtain the specific capacitance and the energy and power densities associated with an electrode material.

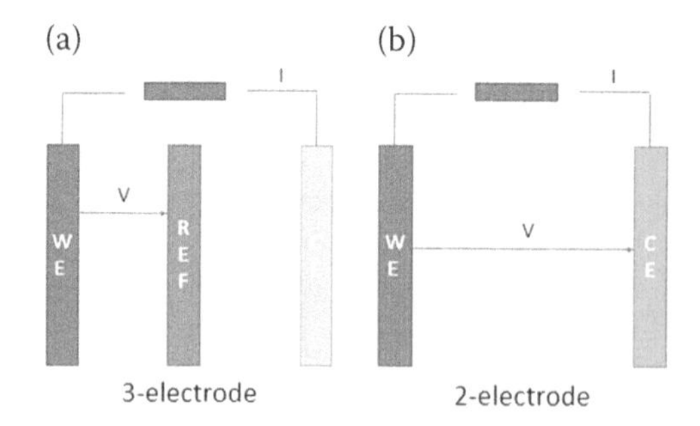

FIGURE 2.2 Schematic diagram for (a) 3 electrode and (b) 2 electrode configurations.

2.3 TECHNIQUES FOR DATA ANALYSIS

2.3.1 Cyclic Voltammetry

Cyclic voltammetry (CV) is a popular electrochemical technique used to investigate the reduction and oxidation processes. Before explaining the measurement, procedures would illustrate the electrochemical cell that is generally employed for carrying out the CV measurements. An electrochemical cell in a three-electrode configuration mainly consists of a working electrode containing the active material, a reference electrode with respect to which the voltage difference is measured, and a counter electrode to pass the current (Figure 2.3a) (Elgrishi et al. 2018). In cyclic voltammetry, the electrode potential ramps linearly with time in a cyclic process. The scan rate (V/s) is the rate of voltage change over time during each cycle. The current is measured between the working and the counter electrodes, and the potential is measured between the working and the reference electrodes. A plot of current (i) vs applied potential (E) is analyzed for the initial forward scan (from time, t_0 to t_1) when an increasingly reduced potential is applied, resulting in the cathodic current to increase initially if reducible analytes are in the system during this potential window. After the reduction potential of the analyte is reached, the cathodic current will decay because of reduced the concentration of reducible analyte. For a reversible redox couple, during the reverse scan (from time, t_1 to t_2), the reduced analyte will re-oxidized. That, in turn, increases the current of opposite polarity (anodic current). Both the oxidation and reduction peaks will be of similar shape for more reversible redox couples. Hence, CV analysis provides the information about electrochemical redox potentials and reaction rates (Elgrishi et al. 2018).

A typical CV plot is represented in Figure 2.3(b), in which potential sweep initiates at −0.4 V for oxidative and more positive potentials. Initially, at smaller potential, no oxidation of the analyte occurs, but exponential increase with the initiation of the oxidization occurs at the working electrode surface with the

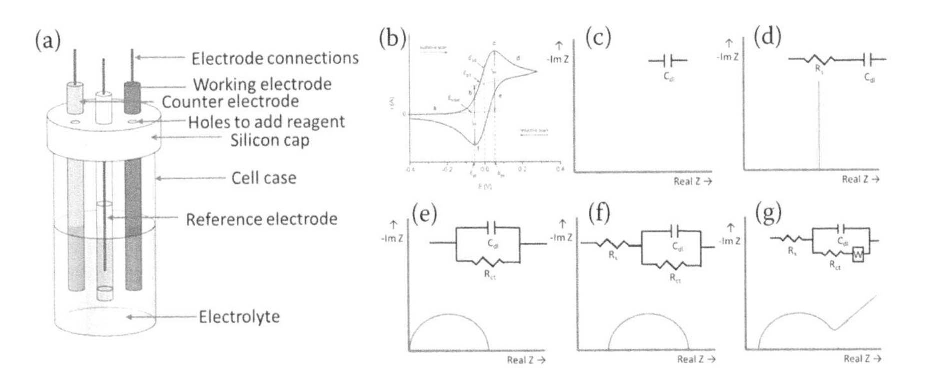

FIGURE 2.3 (a) A schematic illustration of three electrode cell configuration; (b) a representation of typical CV curve; (c–g) Nyquist plot as appears for various ideal systems with their equivalent circuits.

attainment of the onset potential (E_{onset}) of oxidation. In the process of cyclic voltammetry, as the voltage increase, the current increase linearly keeps a constant concentration gradient across the diffuse double layer (DDL) associated with the analyte in the vicinity of the electrode surface (Elgrishi et al. 2018). As the DDL grows, the analyte gets depleted; hence, a gradual loss in the current response from linearity is seen. The current reaches peak maximum, i_{pa} as a result of oxidation when anodic peak potential E_{pa} is reached. Now there can be an increase in the current at more positive potentials by a decay in flux of analyte away from the surface of, the electrode. From this turning point, the current is mainly limited by the mass transport of analyte from the bulk to the DDL interface. This process does not satisfy the Nernst equation because it is slower on the electrochemical timescale; until a steady-state is reached, the current decays as the potential becomes more positive. Finally, a further increase in the potential will no longer have any effect (Elgrishi et al. 2018). Now, for negative potentials (scan reversal), analyte continues to oxidize until the applied potential reaches an optimum value. The oxidized analyte accumulates at the electrode surface and can be re-reduced (Elgrishi et al. 2018).

The reduction and oxidation processes mirror each other, except for an opposite scan direction. For a reversible process, the cathodic and anodic peak currents are required to be of same magnitude but have the opposite sign. Randles-Sevcik equation (2.1) can descried the peak current, i_p, of the reversible redox process. At 298 K, the equation becomes (Ngamchuea et al. 2014):

$$i_p = (2.69 * 10^5) n^{\frac{3}{2}} A C D^{\frac{1}{2}} V^{\frac{1}{2}} \quad (2.1)$$

where 'A' is the electrode area (cm^2), 'D' is the diffusion coefficient (cm^2/s), 'n' is the number of electrons, 'C' the concentration (mol/cm^3), and 'V' the potential scan rate (V/s) (Elgrishi et al. 2018).

2.3.2 Galvanostatic Charging-Discharging

The galvanostatic charge-discharge (GCD) measurement is implemented to monitor the electrochemical capacitance of materials in the current-controlled conditions, while the voltage is noted, unlike the cyclic voltammetry approach. In the field of electrochemical energy-storage devices, GCD becomes one of the largely implemented approaches from a laboratory scale to an industrial scale. The GCD approach is also termed as chronopotentiometry because it measures various parameters, including capacitance cyclability solution and resistance (J. Liu et al. 2018). In this approach, the final potential is monitored with respect to a reference electrode dependent of time on application of a current pulse to the working electrode. The monitored potential varies abruptly on application of current, which is attributed to the loss in IR (internal resistance); after that, there is gradual variation in the concentration of the reactant, attaining zero at the surface of the electrode. The voltage alteration is denoted by the following equation (Elgrishi et al. 2018):

$$V(t) = iR + \frac{t}{C}i(V) \tag{2.2}$$

where 'V' (t) is the voltage as a function of time, 'R' is the resistance, 'C' is the capacitance, and 'i' is the current. The slope of the galvanostatic charge-discharge curve gives the capacitance of the supercapacitor, which will be discussed further in a later section (J. Liu et al. 2018).

2.3.3 Electrochemical Impedance Spectroscopy

Electrochemical impedance spectroscopy (EIS) measurements are carried out with application of alternating-current (AC) frequencies to study various types of resistance offered by the electrode material and its interface with the electrolyte, including interfacial resistance, internal resistance, and charge-transfer resistance (Sunil et al. 2020). Study of data obtained from impedance spectroscopic measurements reveals the data about the interfacial structure, boundaries, and reactions variation. EIS is a sensitive technique and needs lots of care during the measurements. The results are not always clearly understood, attributed to the inadequate mathematical tools related to the equations linking the impedance signal data with the physicochemical parameters. Other methods complimenting EIS can be used to elucidate the interfacial processes perfectly. EIS reveals the direct relation between a real system and an ideal equivalent circuit with discrete electrical components (capacitance (C), resistance (R), and inductance (L)) linked in parallel and series combinations. The graph obtained by plotting real impedance vs. imaginary impedance is termed as Nyquist plot (Sunil et al. 2020). Supercapacitors are the physical assemblies that implement blocking or polarizable electrodes (in case of EDLCs) or electrode materials, which are electroactive (pseudocapacitors). Figure 2.3(c) represents the impedance pattern corresponding to an ideal capacitor comprising a steep rising line that coincides with the imaginary axis of the impedance plot. On the other hand, a physical capacitor contains a shift, resulting in a line parallel to the imaginary axis in a Nyquist plot representing a series arrangement of a series resistance R_s with an electrolytic (Figure 2.3(d)). The simplified Randles cell model includes a solution resistance (or electrolyte resistance), a charge transfer (or polarization) resistance in series, and a double-layer capacitor in parallel. In addition, the simplified Randles cell is used for other more complex models and geometrical capacitance C_g (under the influence of electric field a dielectric polarization of electrolyte). Here, the Nyquist plot would consist of a semicircle only (Figure 2.3(e)). However, as the electrolytic resistance is included in series with a Randles circuit, the semicircle gets shifted, as can be seen in Figure 2.3(f) (Sunil et al. 2020). For real systems, another component, called the Warburg element, has to be considered to account for the diffusion mechanism (Figure 2.3 (g)). At high frequencies, the Warburg impedance is very low and attributed to the lesser distance travelled by the diffusing reactants. However, at the low frequencies, the Warburg impedance become significant, and on the Nyquist plot, it appears as a 45° slope. The place of the impedance spectrum comprising a large capacitance at low frequencies depicts ionic penetration through the entire polymer film or the porous electrode, which is mostly interpreted by the terms of distributed C and R network.

2.4 MODELLING TECHNIQUES FOR SUPERCAPACITORS FROM ELECTRODE TO ELECTROLYTE

A supercapacitor has four main components: electrolyte, electrodes, a membrane separator, and current collectors. On application of potential among the two different electrodes, cations (anions) will get transported to the negatively (positively) charged electrode (Samantara and Ratha 2018a). This charge transport and separation will lead to the formation of electrical double-layers at the interface between the electrode and the electrolyte. With an analogy to the general capacitors, for an electric double-layer, the electrode plays the role of one metal-conducting plate, and electrolytic ions adsorbed acting as counter ions at the electrode/electrolyte interface play the role of other conducting metal plate. Attributed to largely diminished distance between the electrolyte and the electrode, in combination with large surface area of interaction between the electrode and the electrolyte, the charge-storage capacity increases many folds (as the electric field generated becomes very strong) as compared to the conventional capacitors (Samantara and Ratha 2018a). The electrodes used in the supercapacitor generally have high porosity resulting in very high surface area available for interaction with the electrolyte. Selection of electrode and the electrolyte both influences the specific capacitance delivered by the supercapacitor.

As described in an earlier chapter, three classes of electrolytes are mainly used in the supercapacitors. Aqueous electrolytes, including solutions of H_2SO_4 and KOH, are mostly implemented due to their cost-effectiveness and large electrical conductivity. However, the aqueous electrolytes are challenged by low breakdown potential (<1.2 V), along with problems related to corrosion issues (Huo et al. 2019). To mitigate these problems, two kinds of nonaqueous electrolytes are progressively being implemented attributed to their larger working voltage: (1) room-temperature ionic liquids (RTILs), including *N*-methyl-*N*-propylpyrrolidinium bis(trifluoromethanesulfonyl) imide, (2) organic electrolytes like tetraethylammonium tetrafluoroborate in aprotic solvent of acetonitrile. Organic electrolytes can be operated at higher working potential spanning up to 2.7 V (Nasrin et al. 2021). Interestingly, RTILs can be operated at even higher potentials, e.g., 6.0 V. RTILs comprise of entirely organic/inorganic anions and organic cations (Nasrin et al. 2021). Nevertheless, ion-dissolved solvents are main constituents of the organic electrolytes. In addition to the large electrochemical window, RTILs have garnered ample attention in past few years due to their outstanding thermal stability and lower volatility. Various classes of materials have been implemented as electrodes for application in supercapacitors. Particularly, carbon-based materials have been most frequently implemented due to their large specific surface area, lower cost, diversity of form, ease of synthesis, easy processability, and comparatively inert electrochemistry. Carbons in many forms have been designed for implementing in supercapacitors as electrodes, e.g., templated carbon, doped carbon, activated carbon, carbon nanotubes (CNTs), carbide-derived carbons, onion-like carbons (OLCs), and graphene, etc. (Lahrar, Simon, and Merlet 2021). The essence of supercapacitor designing lies in charge-storage characteristics optimization (i.e., improving the energy and power densities) through selecting apt electrodes and electrolytes. In the past few years, heaps of experiments have been dedicated to investigate the capacitive properties of various supercapacitors. In addition to the generally performed electrochemical and physiochemical experiments, molecular simulations suggest the attainable capacitance and structure of electric double-layers in supercapacitors by an investigation of the molecular structures of electrolyte and electrode materials (Lahrar, Simon, and Merlet 2021). This approach is very effective in providing significant understandings of the charge-storage mechanism and can provide explicit assistance toward an ideal model of supercapacitors. In addition to acting as the link between the macroscopic world and microscopic length/time, this kind of simulation can also draw the relation between experiments and theories. The field of supercapacitor modeling has been highly effective in the past few decades, and several publications on the related field publish each month. We will first discuss the fundamental methodology associated with molecular modeling of supercapacitors and thereafter discuss details about the evaluation of experimentally realized parameters. This will be followed by major research results and perceptions. Finally, the possible paths for future works will be discussed.

2.4.1 Basic Methodology

The carbon-based electrodes of a supercapacitor possess highly complex microstructure. Consequently, several kinds of EDLs can be designed for application in a supercapacitor. For instance, some EDLs are created close to the open surfaces; these are planar (e.g., spherical (e.g., OLC), flat graphene sheet, or cylindrical (e.g., CNT)). Some EDLs can also be formed inside thin pores (e.g., slits and cylindrical) (Lahrar, Simon, and Merlet 2021). By contemplating each EDL as a separate capacitor, a supercapacitor can be thought of as several capacitors assembled in series and/or in parallel. This idea makes it realistic to investigate the EDL capacitors to develop insights into the mechanism of charge storage in an actual supercapacitor. Several techniques can be implemented to replicate EDLs. The fundamentals of density functional theory calculations (DFT) and molecular simulations (Lei Zhang et al. 2018) are included the next sections.

2.4.2 Molecular Simulations

Monte Carlo (MC) and molecular dynamics (MD) simulations are well-known simulation techniques applied to the molecules of any system (Lei Zhang et al. 2018). Since EDLs developed in the supercapacitor are molecular phenomenon, these simulation methods are perfectly applicable for modeling supercapacitor. The distinctive benefit of these simulation methods is that they provide data on both the macroscopic properties (e.g., its capacitance) and the microstructure (e.g., ion-density distribution across EDL) (Da Silva et al. 2021). This helps to develop an understanding of the microscopic origins of the capacitance and thus leads to the choice and design of electrode/electrolyte materials. MD simulation is popular as a technique for calculating the equilibrium properties and transport characteristics associated with a classical many-body system (Da Silva et al. 2021). This method is accompanied by numerical solutions of the applicable classical equations of motion to an MD system. The basic algorithm linked with MD is extremely analogous to actual experiments in various ways: The first step is to design a sample experiment; this step is followed by assessing the parameter of interest for a certain duration. The initial condition related to an atomic system, the forces prevailing between the atoms, can be implanted according to the potential of interaction between the atoms. Considering these forces, the position and the velocity of atoms/molecules are revived according to the classical law of motion (Da Silva et al. 2021). The path of the molecules/atoms is examined to obtain microscopic figures of the MD system similar to what is performed in macroscopic experiments (Da Silva et al. 2021). EDLs' simulation of employing MD methods is comparable to recreation of other atomistic trends; literature related to the subject is plentiful. The electrodes in supercapacitors exhibit good electrical conductivity, unlike many other molecular systems, and so their electronic polarizability is at risk of being overlooked. Actually, while executing the experiment, electric potential is precisely controlled instead of the charge on the electrode surface. It is really difficult to carry out modeling of the electrically polarizable bodies to the molecular level; hence, for EDL simulations, the approximate method has been adopted for now, i.e., partial charges are assumed to be distributed uniformly among surface atoms of electrodes. This method is successful for idealized electrodes, including cylindrical, planar, or spherical surface) (Feng, Qiao, and Cummings 2015; Bo et al. 2018). This method, however, is tricky to implement for complex-shaped electrodes in which the charges on different atoms of the electrode differ in the sense time-averaging. For such a case, it is important to focus on a constant electrical potential on the surface of the electrode. Two kinds of approaches have been implemented for this purpose. For the first method, the applied electrical potential is set as a constant on the surface atom of each electrode by altering the charges on the atoms. In the second method, pioneered by the Aluru group of University of Illinois at Urbana-Champaign, with the help of solving a Laplace equation (auxiliary), the potential (electrical) is given on certain points belonging to a numerical grid existing all over the surface of the electrode. The above-mentioned techniques have been applied for shaping nanopores existing in an electrode (Quations, n.d.) (Pean et al. 2016). The setup in detail of the MD model of a electrochemical capacitor associated with electrolytes near the planar electrode can be observed in reference (Bo et al. 2018), and for the case where the nanopores are filled with electrolytes, refer to the publications (Cazade, Hartkamp, and Coasne 2014). Several publications are available where the molecular simulations of EDLs are carried out implementing the MD method and the MC method. Particularly, MC is better suited for modelling the EDLs formed inside the nanopores implementing restricted primitive model (RPM). In such an model, the electrolytic ions are considered as hard spheres possessing a point charge at their center (Banik et al. 2010).

2.4.3 Density Functional Theoretical Calculations

Density functional theory (DFT) is considered a powerful tool for evaluating the equilibrium characteristics in terms of one-body density profiles of many-body systems. As compared to the MD or MC simulations, DFT calculations necessitate fewer computational resources and allow accurate tuning of parameters for modeling, including electrode surface-charge density and ion size and surface potential. For most of the available classical DFT supercapacitor modeling, grand canonical ensemble-based DFT calculations are considered where the electrolyte is denoted by a primitive model and the anions and cations are defined by van der Waals interactions and charged hard spheres refraining from the complex ionic structure. A supercapacitor DFT modeling possessing nanopores filled with room temperature ionic liquids can be found in the reference (Kui Xu et al. 2020).

2.4.4 Capacitance Calculations

After determining the ion-density distribution within a modeled supercapacitor (or EDL) after implementing DFT calculations or molecular simulations, a supercapacitor's macroscopic characteristics can be calculated (Kui Xu et al. 2020). The most significant characteristic of a supercapacitor is its specific capacitance. The integral and differential capacitances of a supercapacitor are defined as

$$C_{int} = \frac{\sigma}{(\varnothing_{EDL} - PZC)} \tag{2.3}$$

$$C_d = \frac{d\sigma}{(\varnothing_{EDL}\ C_d)} = \frac{d\sigma}{\varnothing_{EDL}} \tag{2.4}$$

where C_d and C_{int} represents the differential and integral capacitances, respectively; σ represents surface charge density of the electrode; ϕ_{EDL} is the potential gap between electrolyte and the surface of the electrode. In case σ is not precisely imposed in MD simulations, it can be entered as (Feng et al. 2011):

$$\sigma = -\int \varnothing_{EDL}(u/R)^N \rho_e du \tag{2.5}$$

where ρ_e represents the space-charge density measured for an EDL obtained from DFT calculations or molecular simulations, 'u' is directed along the normal direction of electrode surface, 'R' is the radius of the spherical or cylindrical electrode, N = 0 is for the planar electrode, and N = 1 and 2 are for cylindrical and spherical electrodes, respectively. Meanwhile, ϕ_{EDL} can be calculated from the potential distribution, ϕ, which can be obtained by solving Poisson equation as follows:

$$\nabla^2 \varnothing = -\rho_e / \varepsilon_r \varepsilon_0 \tag{2.6}$$

where ε_r and ε_0 represents the dielectric constants of the EDL and the vacuum permittivity, respectively. Particularly, for EDLs in the vicinity of the planar electrodes, the potential distribution can be represented by:

$$\varnothing(z) = -1/\varepsilon_r\varepsilon_0 \int_z^0 (z - u)\rho_e(u)du - \sigma\varepsilon_r\varepsilon_0 z \tag{2.7}$$

For cylindrical electrodes,

$$\varnothing(r) = -1/\varepsilon_r\varepsilon_0 \int_r^R u\rho_e(u)\ln(r/u)du - (\sigma R/\varepsilon_r\varepsilon_0)\ln(r/R) \tag{2.8}$$

For spherical electrodes,

$$\varnothing(r) = -1/\varepsilon_r\varepsilon_0 \int_r^R (1 - u/r)u\rho_e(u)du - (\sigma R/\varepsilon_r\varepsilon_0)(1 - R/r) \tag{2.9}$$

Most present-day simulations related to the supercapacitors refrain from considering the double layers formed inside electrodes. Those kinds of double-layers indeed exist and can influence the capacitance of a supercapacitor. Modelling the double-layers existing inside the electrodes and at electrode/electrolyte interfaces at the same time, however, is very difficult. Such simulations will differentiate the electronic and atomic degrees of freedoms of these EDLs. This differentiation can be achieved in principle through integration of MD and DFT approaches. Specifically, the statistical dissemination of electrolytic ions/solvent molecules can be determined by MD, and the calculations related to the electrodes can be completed using DFT. Nevertheless, this combination is computationally highly intensive. Furthermore, several published modeling studies demonstrate that not considering the contribution of the double-layer within electrode and the electronic degree of freedom does not severely affect the prediction of EDLs when implemented with nonaqueous electrolytes. Further, the next section elaborates the parametric studies of a supercapacitor from the experimental results.

2.4.4.1 Calculation of Specific Capacitance from CV, GCD, and EIS

The specific capacitance of a supercapacitor can be calculated both from the CV as well as from the GCD plots, as shown in Figure 2.4 (Kumar and Misra 2021; Kumar et al. 2020; J. Liu et al. 2018).

- The specific capacitance from the CV can be calculated as follows:

$$\mathrm{C_{sp}} = \frac{\int \mathrm{IdV}}{\mathrm{m\Delta V}} \tag{2.10}$$

where, m is the total mass of the electrode active material, $\int \mathrm{IdV}$ is the area enclosed by the CV plot, and ΔV is the working potential. The specific

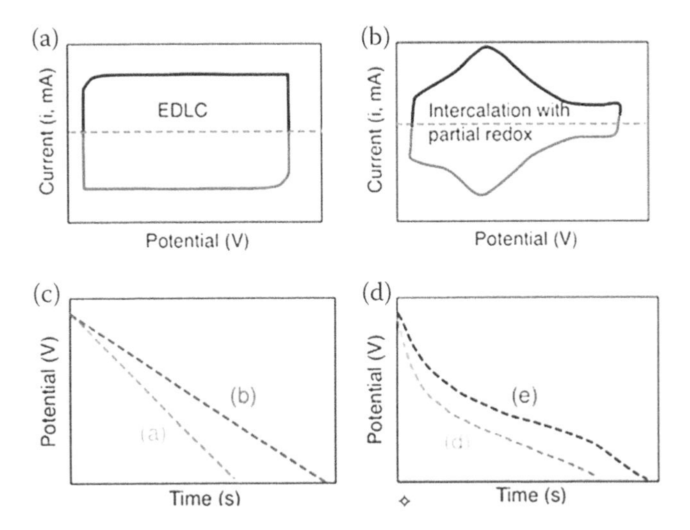

FIGURE 2.4 The typical cyclic voltammogram and galvanostatic charging discharging plots for the electric double-layer type material and pseudocapacitive type material.

Source: Reprinted with permission from (Gogotsi and Penner 2018), Copyright © 2018 American Chemical Society.

capacitance can be calculated from mass, area, or volume of the electrodes in the device.

- It can also be calculated from GCD curve using the discharging time as:

$$C_{sp} = \frac{I\Delta t}{m\Delta V} \tag{2.11}$$

where, I is the applied current, Δt is the discharging time, m is the total mass of the electrode active material, and ΔV is the working potential.

However, equation 2.11 can be avoided to calculate for a GCD curve when the slope of the discharge curve changes with time and is not constant.

Therefore, the specific capacitance for such materials should be calculated, as given below: (Roldán et al. 2015)

$$C_{sp} = \frac{2I\int Vdt}{m \times V^2\Big|_{V_i}^{V_f}} \tag{2.12}$$

- To calculate the specific capacitance from the EIS, the following equation should be used:

$$C_{sp} = \frac{1}{m2\pi fZ''} \tag{2.13}$$

where, f is the frequency and Z" is the imaginary part of the impedance obtained at the given frequency.

2.4.4.2 Why We Cannot Calculate the Specific Capacitance of a Battery-like Material

Typical CV curves for pure EDLC and pseudocapacitive response are shown in Figure 2.4. As can be seen in the Figure 2.5, for a battery-type material, the capacitance remains zero for a significant potential range above 3.4 V and below 3.2 V, while in the middle, i.e., at the potential range from 3.2 V to 3.4 V (0.2 V) 1866 F g^{-1} is the value of specific capacitance is (Figure 2.5). Now, if we calculate the energy density with this capacitance value using the equation:

$$E = \frac{1}{2}CV^2 \tag{2.14}$$

Where C is the specific capacitance and V is the potential range, the value of the energy density calculated will be as high as 1019 Wh kg^{-1}. However this apparent value is overestimated because the value of specific capacitance is 1866 F g^{-1} in the working potential of 0.2 V, which will result in an energy density of value 10.33 Wh kg^{-1} (10 times smaller as compared to the previous case). So, for battery-type material instead of specific capacitance the specific capacity is measured.

2.4.4.3 The Contribution of Charge: Diffusion-Controlled and Capacitive

Area enclosed by a cyclic voltammogram indicates the total stored charge originating from non-faradaic and Faradaic reactions. The total charge stored can be classified into three components: non-Faradaic contribution originated from electric double-layer capacitance, the diffusion-controlled Faradaic contribution, and the non-diffusive Faradaic contribution that occurs from the charge-transfer reaction with the surface-bound atoms, often referred to as pseudocapacitance

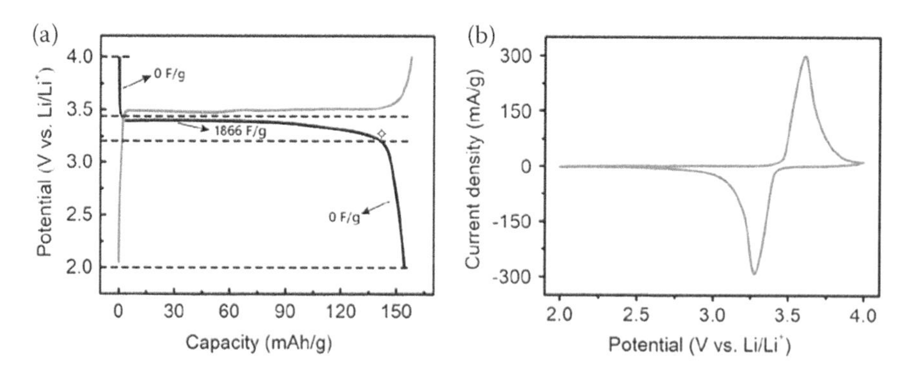

FIGURE 2.5 The typical (a) charging and discharging plot and (b) CV plot for a battery-type material.

Source: Reprinted with permission from (Mathis et al. 2019), Copyright © 2019, John Wiley and Sons.

(John Wang et al. 2007; Brezesinski et al. 2010). The current i from the CV curve obeys the power law, as shown in equations (2.15) and (2.16) (Lindström et al. 1997; John Wang et al. 2007).

$$i = a\nu^{b} \tag{2.15}$$

$$\log(i) = b\log(\nu) + \log a \tag{2.16}$$

where a, b are adjustable parameters and υ is the scan rate. The value of parameter b can be obtained from the slope of the linear plot of log i vs log υ. The obtained b values can classify the charge-storage mechanism. Generally, diffusion-controlled processes possess $b = 0.5$ satisfying Cottrell's equation: $i = a\nu^{1/2}$ and $b = 1$ for capacitive dominated processes (John Wang et al. 2007).

Furthermore, a quantitative separation of capacitive and diffusion-controlled contribution from scan rate dependent voltammetric current is introduced by Dunn and co-workers. The current response i at each voltage (V), is the sum of the two contributions, as expressed in the following equation: (John Wang et al. 2007):

$$i(V) = k_1\upsilon + k_2 v^{1/2} \tag{2.17}$$

where k_1 and k_2 are constants. In this equation, $k_1\upsilon$ represents the surface capacitive effects while $k_2 v^{1/2}$ represents diffusion-controlled processes. Dividing the square root of the scan rate on both sides of the equation for analytical purposes yields:

$$i(V)/\upsilon^{1/2} = k_1\upsilon^{1/2} + k_2 \tag{2.18}$$

Therefore, after plotting $i(V)/\upsilon^{1/2}$ vs $\upsilon^{1/2}$ at a given voltage, the constants k_1 and k_2 can be obtained. Thus, by determining k_1 and k_2, the current response can be quantified due to surface capacitive and diffusion-controlled processes at a specific potential. This characterization technique has been used to evaluate the electrochemical performance of the electrode materials, as previously reported and shown in Figure 2.6.

A similar kind of analysis was first developed by Trasatti et al. (S ARDIZZONE, G FREGONARA 1990) to separate the surface-capacitive effect from the diffusion-controlled process using the relation between capacitance and the scan rates. Capacitance due to surface reactions will be independent of the variation of scan rate. On the other hand, the semi-infinite linear diffusion-controlled capacitance will vary with $\upsilon^{1/2}$. The total stored charge (q_T) can be demonstrated as the charge at 0 mV/s because there is sufficient time for every reaction to occur, while the charge stored at the surface (q_S) can be demonstrated as the charge stored at an infinite scan rate. The difference between these two charges ($q_T - q_S$) will provide the stored charge due to diffusion-controlled processes. The total

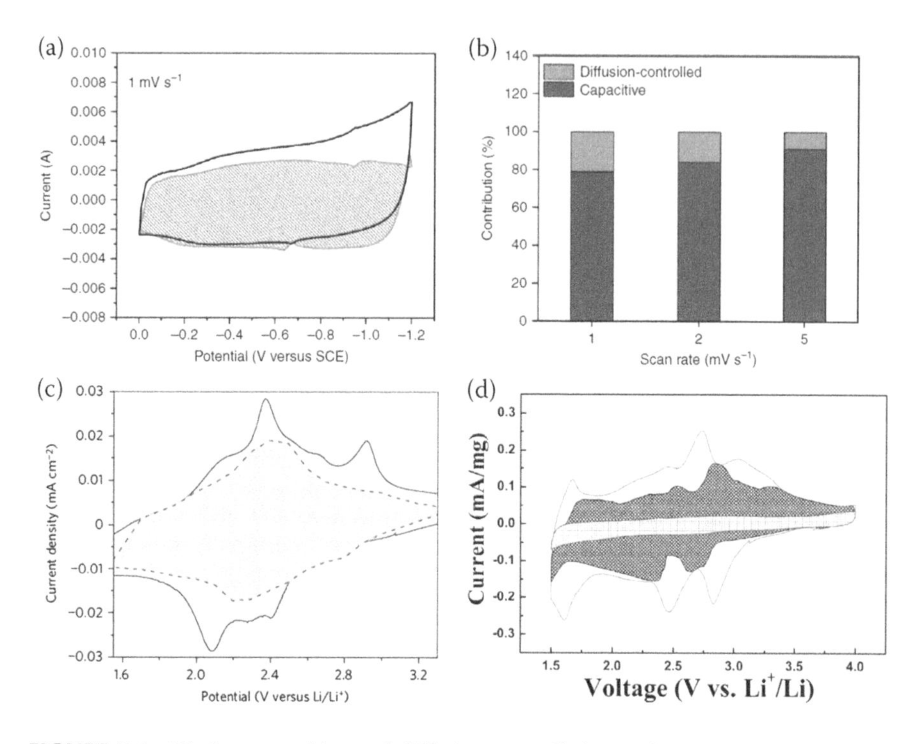

FIGURE 2.6 Distinct capacitive and diffusion-controlled capacitance resulting in charge storage in a supercapacitor. (a) Voltammetric response at 1 mVs^{-1}. The capacitive charge-storage corresponding to the entire current is indicated by the shaded region. (b) The capacitance contribution at different scan rates (1, 2 and 5 mVs^{-1}). a,b Reprinted with permission from (Owusu et al. 2017), Copyright © 2017 Springer Nature. (c) Another example of cyclic voltametric response from mesoporous MoO_3 film at 0.1 mVs^{-1}. The capacitive charge-storage corresponding to the entire current is indicated by the shaded region. Reprinted with permission from (Brezesinski et al. 2010), Copyright © 2010, Springer Nature. (d) Charge storage mechanism in V_2O_5/CNT electrodes at 0.1 mV^{-1}. The grey shaded portion corresponds to capacitive contribution.

Source: Reprinted with permission from (Sathiya et al. 2011), Copyright © 2011, American Chemical Society.

voltammetric charge, q_T, can be obtained by plotting the inverse of the voltammetric charges (q) against the square root of scan rates assuming semi-infinite linear diffusion of ions, as expressed in the following equation:

$$\frac{1}{q} = \text{constant}\cdot \nu^{1/2} + \frac{1}{q_T} \tag{2.19}$$

Internal resistance of the electrodes and nonconformity from semi-infinite linear diffusion of ions data points at higher scan rates may deviate from linearity and those deviated points were avoided during the linear fitting. The "total stored

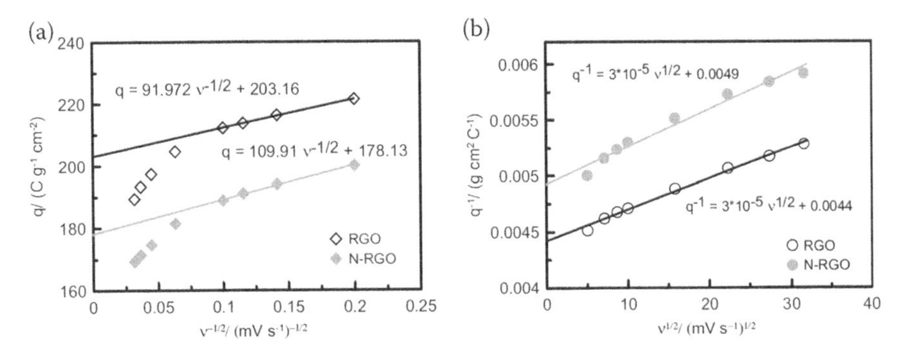

FIGURE 2.7 (a) Dependence of q on $\upsilon^{-1/2}$ and (b) dependence of $\frac{1}{q}$ on $\upsilon^{1/2}$ for RGO and N-RGO in 1 M KOH.

Source: Reprinted with permission from (Y. H. Lee, Chang, and Hu 2013), Copyright © 2012, Elsevier.

charge" is equal to the sum of the charges due to surface reactions and diffusion-controlled reactions. Plotting the voltammetric charge, q, vs the inverse of the square root of the scan rates should yield a linear plot as presented in the following equation (assuming semi-infinite linear diffusion of ions):

$$q = \text{constant}\cdot v^{-1/2} + \frac{1}{q_0} \tag{2.20}$$

where q_0 can be defined as so-called stored charge at the outer surface of the electrode or the stored charge due to the capacitive reactions at the surfaces. After linear fitting of the plot and extrapolating the fitted line to the y-axis will provide q_0. The difference between the calculated q_T and q_0 yields the maximum diffusion-controlled capacitance. The above characterization technique has been used to evaluate the electrochemical performance of the electrode materials as previously reported and shown in Figure 2.7.

3 State-of-Art Supercapacitor Design

3.1 FUNDAMENTALS OF PLANAR SUPERCAPACITOR

Rapid miniaturization of the electronic devices has stimulated a demand for the compatible-energy storage system, and thus, planar or micro supercapacitors (MSCs) have garnered huge attention. As compared to the conventional bulk supercapacitors associated with either a stack geometry or bulk porous materials, the main characteristics of planar MSCs are a rendered thinness with much-reduced mass to the complete device. Moreover, MSCs have a smaller device size and provide flexibility to be implanted as an arbitrary substrate, including plastics. The small interspaces between the electrodes allow the ions from the electrolyte to get transferred rapidly, resulting in a ultrahigh power delivery due to the reduced diffusion distance (El-Kady and Kaner 2013). The separator mediated-ion transport from the electrolyte to the electrodes in conventional supercapacitors is now unnecessary in that kind of MSC. The tunable thickness associated with the planar MSCs makes it suitable for several types of fabrication of miniaturized devices for direct integration with electronics on an industrial scale (El-Kady and Kaner 2013). Unlike the planar supercapacitor, the supercapacitor with stack configuration is not suitable for the transfer of gel or solid electrolytes, consequently causing a significant reduction in the power delivered. Particularly, the association of the traditional bulk supercapacitors with the modern electronic gadgets results in a big challenge that necessitates several steps related to processing the arrangement of the layers of active materials, current collectors, and electrolytes, and to connect all the components at the ends (Pech et al. 2013). Henceforth, MSCs possess advantages of ease in designing electrically separated electrodes and integrating them into small dimension electronics existing on a similar plane. Planar supercapacitors also allow broad suitability for several kinds of electrolytes and the simple design alternation in patterns of the microelectrodes. Thus, this arrangement results in a decrease in the internal impedance due to the loss of distance between adjacent interdigital electrodes. To start with, in 2003, Sung et al. were responsible for the primary fabrication of the first planar MSCs prototype in combination with a liquid-electrolyte dropped on a substrate made up of silicon. This work contributed greatly to the fabrication and design of flexible and all-solid-state MSCs (Sung et al. 2006). This research is a pioneering example of planar MSCs, where an interdigital MSC was fabricated on Au and Pt metals-based microelectrode array using conducting polymer. The deposition techniques used in this method were photolithography and electrochemical polymerization. In this work, conducting polymers, poly-(3-phenylthiophene) (PPT), and polypyrrole (PPy) were grown on the electrodes in micro scale electrochemically. So procured MSC comprised of 50 fingers was assembled with two electrolytes *viz.* non-aqueous (0.5 M Et_4NBF_4 in acetonitrile) and aq. (0.1 M H_3PO_4) and together with the cell potentials recorded varying between 0.6 and 1.4 V (Sung, Kim, and Lee 2003). Since these MSC devices used liquid electrolytes, in practical applications, leakage of electrolytes must be considered. Therefore, the designing of all-solid-state MSCs that will implement only solid materials is a mandate. In 2004, Sung and co-workers designed gel-polymer electrolytes integrated all-solid-state MSCs on SiO_2/Si substrate through photolithography, solution-casting, and electrochemical polymerization. Gold microelectrodes acted as the current collectors, and PPy played the role of electrochemically active material. An aqueous-based PVA/H_3PO_4 electrolyte, two kinds of gel-polymer electrolytes, and a non-aqueous polyacrylonitrile (PAN)/$LiCF_3SO_3$-EC/PC electrolyte were tested. Tuning of the capacitance of the device was possible by controlling the content of deposited PPy. The working potential window achieved was ~0.6 V. They concluded that the device performance of all-solid-state MSCs mirror the performance of MSCs associated with a liquid electrolyte (Sung, Kim, and Lee 2004). After 2006, aggressively increasing attention has been devoted to developing planar supercapacitors concentrating on the synthesis of efficient nanostructured materials for the electrodes and utilizing techniques such as thin-film deposition, including inkjet printing (Pech et al. 2017), electrochemical polymerization (Sung et al. 2006), and layer-by-layer assembly (Beidaghi and Wang 2012), together with the creation of novel device structures from two-dimensional to three-dimensional microelectrodes (Shen et al. 2011). In conclusion, all the design parameters in MSCs, along with the suitable materials' choice, aim to significantly enhance the power and energy densities, along with the cyclic stability and frequency response.

3.2 ELECTRODE DESIGN

Considering the cumbersomeness of the generally implemented microfabrication methods, the low-cost micro-devices for making micropatterns on substrates are desirable for a wider range of applications. ElKady and Kaner reported a scalable fabrication method of MSC by direct laser writing on graphene oxide (GO) films, implementing a standard light scribe DVD burner (El-Kady and Kaner 2013). This approach is a facile, cost-effective, high-throughput lithographic technique and refrains from the use of masks and extra complex or operations processing. Figure 3.1 presents the designing MSCs involving laser-scribed graphene (LSG).

Primarily, in this process, a disc grown with GO film on it was put into a light scribe DVD, equipped with a computer-designed circuit created with a laser to create patterns on the GO film. An electrolyte was used to procure a planar LSG-MSC. This approach can be easily increased in scale for the fabrication of >100 MSCs for solid micro-devices on a flexible substrate within 30 minutes or less. A wide potential range of 2.5 V was obtained with the fabricated MSCs

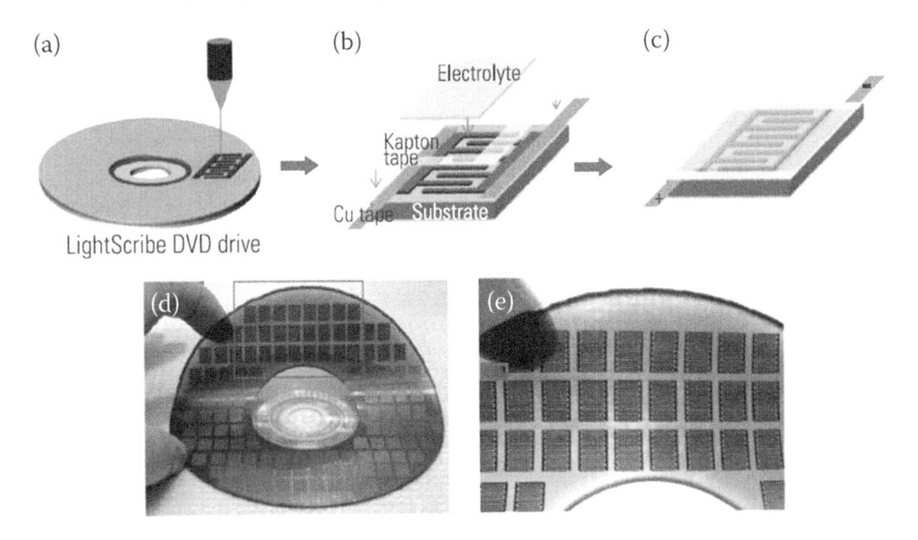

FIGURE 3.1 (a–c) The fabrication procedure for LSG-MSCs. (d,e) A single disc with direct writing of more than 100 LSG-MSCs.

Source: Reprinted with permission from EI-Kady and Kaner[40]. (Copyright 2013, Macmillan Publishers Limited).

with 1-butyl-3-methylimidazolium bis(trifluoromethylsulfonyl)imide (an ionogel electrolyte) fumed with silica. This MSC presented an augmented charge-storage capacity with a power density of ~200 W cm^{-3} along with a rational rate capability offering a time constant of 19 ms. Even though graphene-based MSCs presented a vitally improved performance in thin-film device architectures using easily accessible manufacturing technologies, but the power density of the MSC is lower compared to the thin-film batteries based on lithium. Further, Wu et al. reported planar interdigitate graphene-based MSCs by incorporating graphene (MPG) films reduced by methane plasma deposited on any arbitrary substrates (Figure 3.2) (Z. S. Wu et al. 2013).

MPG films showed a high electrical conductivity of ~345 S cm^{-1} and, using the planar geometry of graphene-based MSCs with H_2SO_4/PVA, could exhibit a volumetric capacitance of ~17.9 F cm^{-3}, an areal capacitance of ~80.7 μF cm^{-2}, together with the energy density of 2.5 mWh cm^{-3} and a power density of 495 W cm^{-3}. Furthermore, these devices also exhibited a better cycling stability corresponding to ~98.3% capacitance retention at the scan rate of 50 V s^{-1} following 100 000 cycles. These micro supercapacitors could operate at very high scan rates with values attaining up to 1000 V s^{-1}, which is three orders of magnitude greater compared to conventional supercapacitors. Thus, the benefits of planar architecture of MSCs are mentioned over the bulk classical stack with sandwich-like architecture supercapacitors. These MSCs can be integrated with other energy-storage devices by connecting in parallel or in series combination to satisfy some particular applications requiring larger operating currents or applying voltages in a short time. Thus, the MSCs based on graphene having a planar geometry provide a ocnsiderable benefit for various flexible and miniaturized electronic applications.

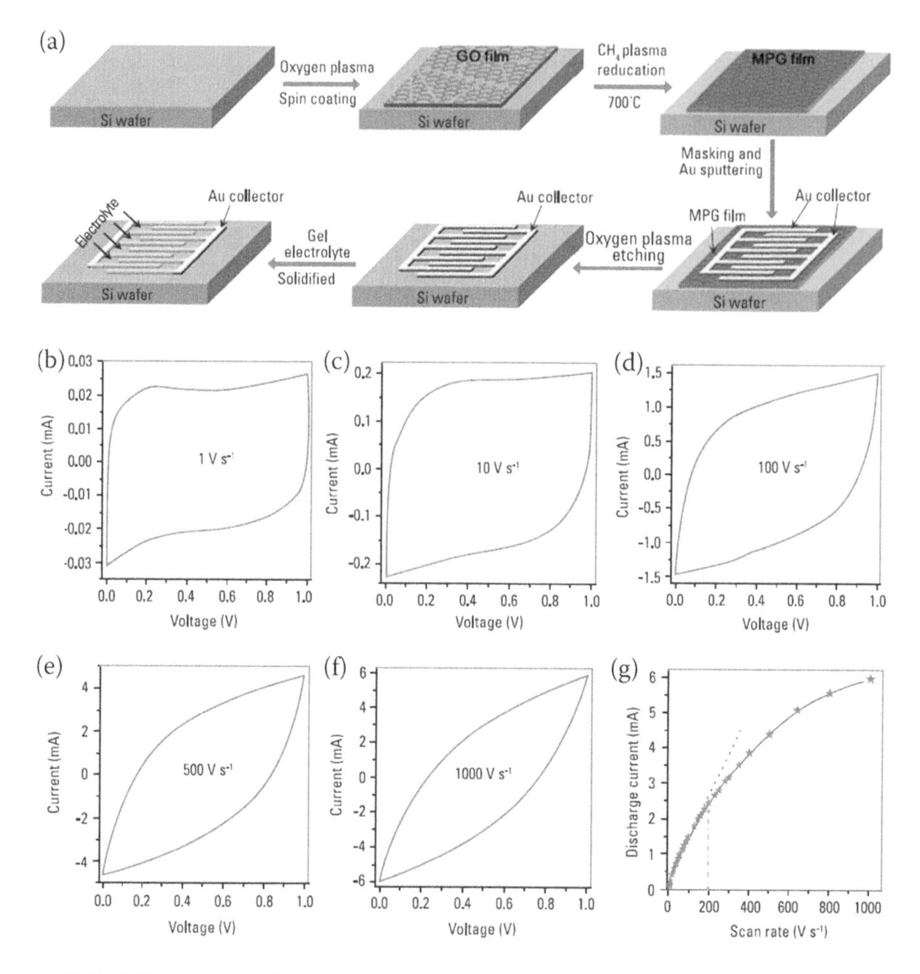

FIGURE 3.2 (a) All-solid-state interdigitated graphene-based MSCs grown on silicon wafer has been shown schematically. (b–f) Cyclic voltammograms of the MSCs at 1 to 1000 V s^{-1}. (g) Discharge current plotting with the scan rate.

Source: Reprinted with permission from Wu et al. (Z. S. Wu et al. 2013). (Copyright 2013, Macmillan Publishers Limited.)

3.3 MATERIALS DESIGN

Implementing atomically thin 2D graphene, novel architectures for the thin-film planar MSCs has been reported. These MSCs utilize the atomically thin layer of graphene with flat morphology. In addition, MSCs due to the planar surfaces are suitable for a short ion diffusion from the electrolytic by interacting along the entire graphene sheet surface, as shown in Figure 3.3 (J. J. Yoo et al. 2011).

Yoo and research group fabricated ultrathin planar supercapacitors incorporating both multilayer-reduced graphene oxide (rGO) film and pristine CVD-grown monolayer graphene. The planar supercapacitor exhibited a significant enhancement in capacitance

(a)

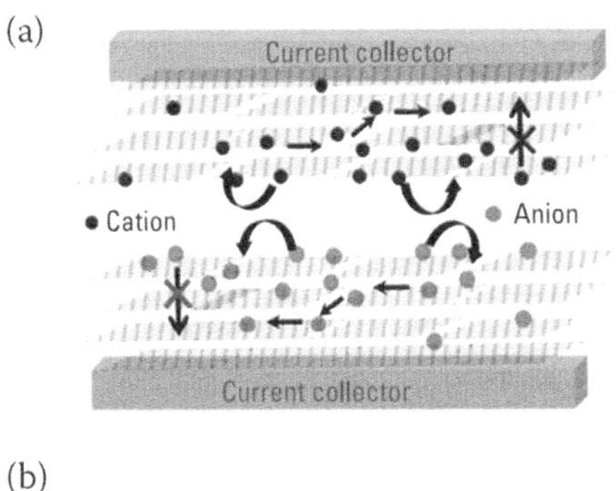

(b)

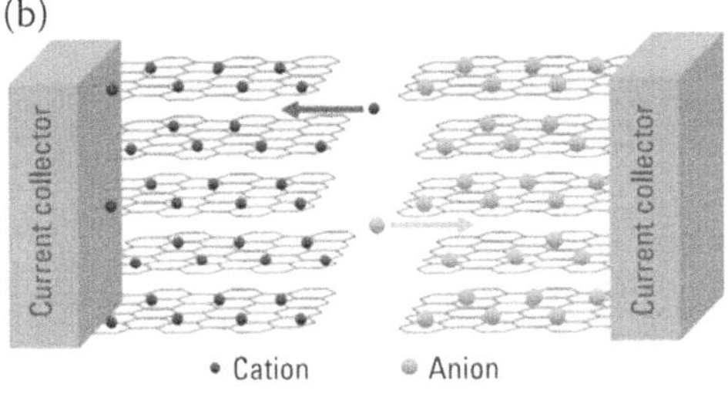

FIGURE 3.3 Supercapacitor device design, where graphene is arranged (a) stacked and (b) planar geometries.

Source: Reprinted with permission from Yoo et al. (J. J. Yoo et al. 2011). (Copyright 2010, American Chemical Society.)

compared to the conventional supercapacitors. The synthesized planar devices delivered a specific capacitance corresponding to a value of ~80 μF cm^{-2} for the monolayer graphene film. The multilayer rGO film exhibited a specific capacitance of 394 μF cm^{-2}. Graphene-based MSCs were patterned through hydrated graphite oxide (GO) films' laser reduction (Gao et al. 2011) by Ajayan's group. In this process, substantial quantity of trapped water in GO acts as a good ionic conductor and also an insulator (electrical) for an electrolyte, as well as an electrode separator for ion transport. Both sandwich-like and planar supercapacitors were planned in several shapes and patterns (Figure 3.4), which were directly grown on a GO paper.

A concentric circular geometry of the planar supercapacitor exhibited an areal capacitance of~0.51 mF cm^{-2}, which was almost double that of the sandwich supercapacitor and showed enhanced cyclic stability to a ~35% reduction of capacitance following 10,000 cycles. However, the long gap between the electrodes of the devices could result in lowered frequency response, high internal resistance (6.5 k), and a poor rate performance. However, this technique of large-scale production showed a promise for future developments. A scalable and facile method to manufacture graphene-cellulose membrane paper (GCP) is designed by Weng et al., which will act as binder-free freestanding flexible electrodes for MSCs (Figure 3.5) (Weng et al. 2011). The electrode corresponding to flexible GCP electrode comprises an interwoven 3D structure of cellulose fibers and graphene sheets. 120 F g^{-1} of specific capacitance was attained by the GCP electrode (in graphene weight) and >99% of capacitance retention over 5000 cycles. A flexible interdigitated MSC with the electrode

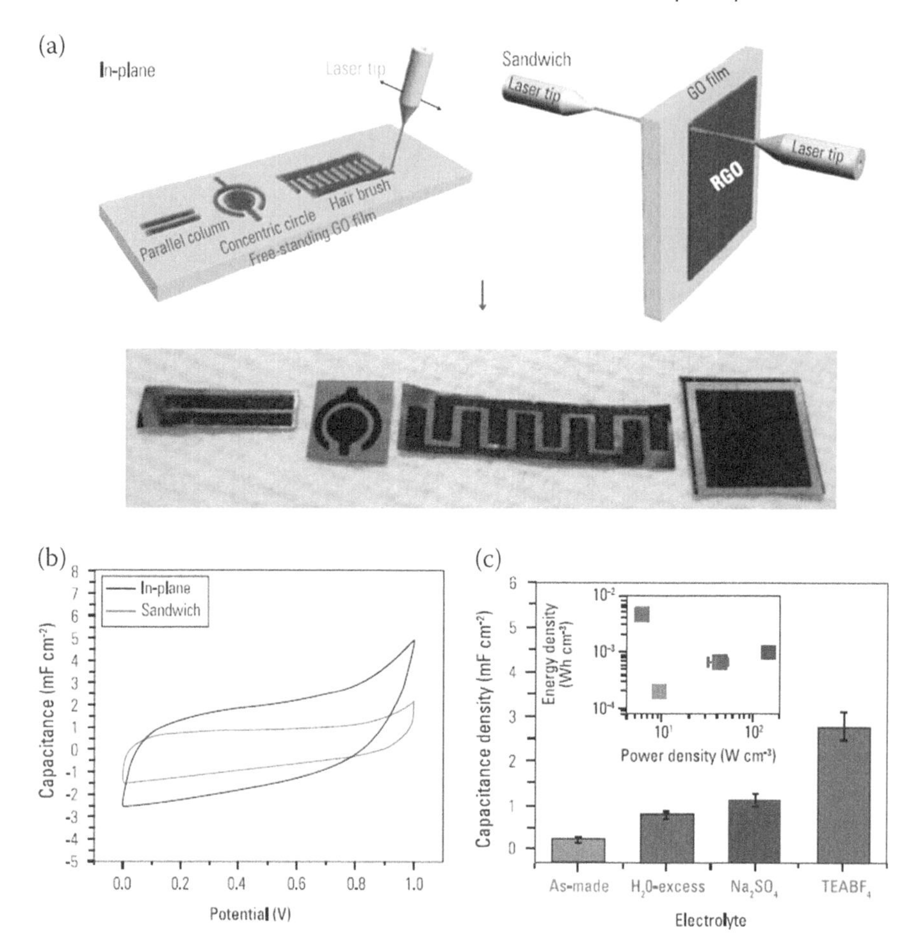

FIGURE 3.4 (a) Laser-patterning of hydrated GO films in combination with a picture of the patterned films in various geometries. (b) Cyclic voltammograms of both kind of devices at the scan rate of 40 mV s^{-1}. (c) Areal capacitance obtained for the sandwich-type MSC device with various electrolytes and Ragone plot for the different fabricated devices (Inset).

Source: Reprinted with permission from Gao et al.(Gao et al. 2011) (Copyright 2011, Macmillan Publishers Limited.)

material of GCP was designed involving H_2SO_4/PVA gel electrolyte on a polydimethylsiloxane substrate. So-obtained MSC exhibited an areal capacitance of ~7.6 mF cm^{-2}. Xu's group designed a ultrathin, flexible, and all-solid-state MSCs based on graphene having H_3PO_4/PVA gel as the electrolyte. Very thin rGO-based interdigital electrodes were fabricated on a polyethylene terephthalate (PET) substrate with the help of combination of photolithography and electrophoretic deposition (Figure 3.6) (Niu, Zhang, et al. 2013).

A high specific capacitance of 286 F g^{-1} was resulted due to the small ion diffusion pathway, in rGO-MSCs. This value was three times improved as compared to the traditional rGO-based supercapacitors having a ~86 F g^{-1} of specific

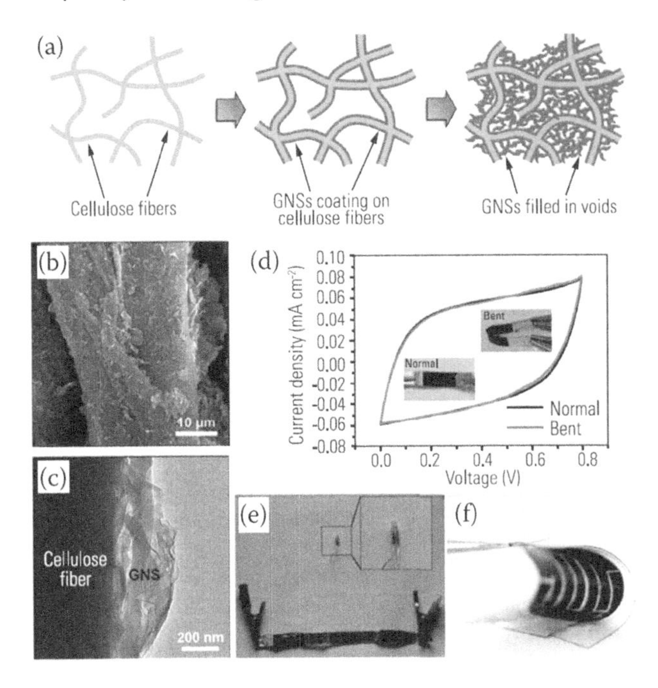

FIGURE 3.5 (a) Structural of initial and final form of GCP. GCP membrane shown by (b) SEM and (c) TEM images. (d) CV curves monitoring the responses in normal and bent states of MSC. (e) MSC device powered red LED. (f) MSC comprising of a flexible GCP.

Source: Reprinted with permission from Weng *et al.* (Weng et al. 2011). (Copyright 2011, John Wiley & Sons, Inc.)

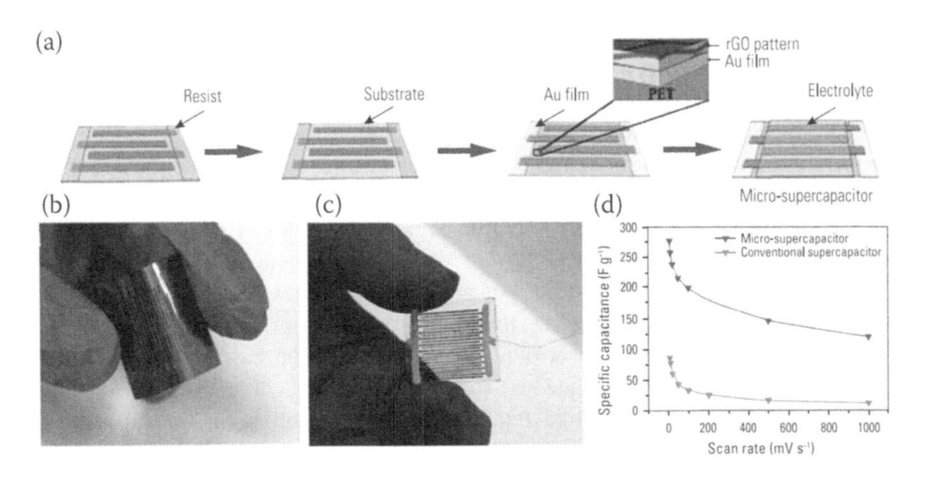

FIGURE 3.6 (a) Fabrication of rGO-MSCs by using both the photolithography and electrophoresis. Digital pictures of (b) rGO pattern and (c) microsupercapaccitor on a PET substrate. (d) Comparison of specific capacitance associated with MSCs based on rGO and traditional supercapacitors at various scan rates.

Source: Reprinted with permission from Niu et al. (Niu, Zhang, et al. 2013). (Copyright 2013, John Wiley & Sons, Inc.)

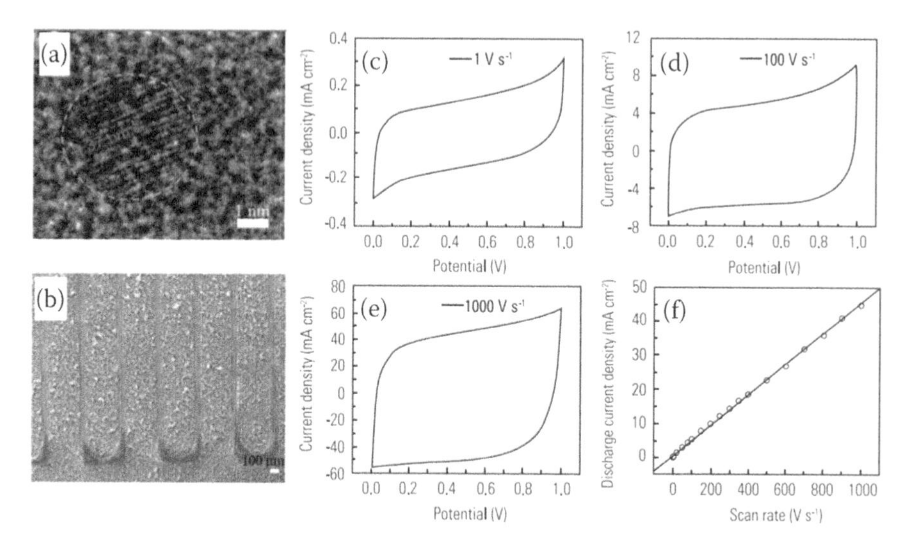

FIGURE 3.7 (a) TEM micrograph of a single GQD. (b) SEM micrograph of electrodes interdigitated based on GQDs. (c–e) CV curves at (c) 1, (d) 100 and (e) 1000 V s^{-1}. (f) Change in current density with scan rate.

Source: Reprinted with permission from Liu et al. (W. W. Liu et al. 2013). (Copyright 2013, John Wiley & Sons, Inc.)

capacitance. Quantum dots (GQDs) based on graphene show exclusive physical and chemical characteristics, large electrical conductivity, adequate edge defects, easy functionalization, and chemical stability, rendering them promising for a supercapacitor (Shinde and Pillai 2013). Yan's group used electrode materials based on GQDs for application in MSCs (GQD-MSCs) (W. W. Liu et al. 2013) tested in ionic liquid ($EMIMBF_4$) and 0.5 M Na_2SO_4 electrolytes. MSCs were designed on interdigital Au microelectrodes by the electrodeposition of GQDs (Figure 3.7).

GQD-MSCs could function at a scan rate as high as 1000 V s^{-1} and grasp rapid frequency response corresponding to a 103.6 μs of time constant when tested in an aqueous electrolyte and 53.8 μs when tested in an ionic liquid electrolyte associated with a capacitance retention of ~97.8% over 5000 cycles. The improved characteristics of MSCs based on GQD are due to the beneficial characteristics of GQDs, such as an adequate number of active surface sites, large specific surface area, and edges providing adequate interfaces for rapid ion adsorption/desorption. To attain high-energy and power densities, much attention has been given to the asymmetric supercapacitors (aqueous) consisting of a capacitive electrode for high power density and a battery-like Faradic electrode, resulting in the increase in operating voltage (Z. S. Wu et al. 2010). Yan's group projected designing of a GQD//MnO_2 asymmetric MSCs, applying MnO_2 nanoneedles playing the role of the positive electrode and the negative electrode comprising of GQDs equipped with 0.5 M Na_2SO_4 (W. W. Liu et al. 2013). The asymmetric MSCs consisting of GQD//MnO_2 were designed using a two-step electrodeposition. Initially, electrodeposition of GQDs was performed on one face of the interdigital Au electrodes in

50 mL dimethylformamide (DMF) solution engaging 6.0 mg $Mg(NO_3)_2{\cdot}6H_2O$ and 3.0 mg GQDs at 80 V for 30 min. This step was followed by an MnO_2 electrochemical deposition on another side of the Au electrodes within 0.1 M $NaNO_3$ and 0.02 M $Mn(NO_3)_2$ for 5 minutes, with an operating potential between –1.2 and 1.2 V at 1 mA cm^{-2}. The specific capacitance obtained for the GQD//MnO_2 asymmetric MSCs was ~1107.4 μF cm^{-2} and the energy density corresponded to a value of ~0.154 μWh cm^{-2} which was two times improved over the symmetric GQD-MSCs in Na_2SO_4 electrolyte. Following a similar approach, they also designed an all-solid-state MSC asymmetric in nature equipped with PVA/H_3PO_4 gel electrolyte through the electrodeposition of polyaniline (PANI) nanofiber as a positive electrode and GQDs playing the role of negative electrode on both the side of Au microelectrode [64]. The so designed asymmetric MSCs comprising of GQD// PANI delivered a good rate capability, a scan rate of 700 V s^{-1} in combination with ~85.6% of capacitance retention followed by 1500 cycles in combination with a small time constant of 115.9 μs when tested in a solid-state electrolyte.

3.3.1 Graphene Carbon Nanotube Hybrid

As mentioned previously, the planar architecture can lead to an enhancement in the accessibility of ions up to the graphene sheet surface and hence enhance the charge-storage ability. Moreover, it is normal to expect that the charge-storage capability and energy density of planar MSCs is possible to improve in the presence of additional capacitive spacers, including CNTs, electrically conductive polymers, and nano-scaled metal oxide within the gap between the graphene sheets. The implementation of such capacitive spacers not only results in limiting the graphene sheet agglomeration and restacking along with enhancing the available surface for charge storage. Apart from that, the synergetic effect developed in between the graphene and the spacers leads to improvement in the device performance (Z. S. Wu et al. 2012). Wang's group implemented photolithography lift-off followed by the electrostatic spray deposition for making the interdigital microelectrodes (width of100 μm and interspace of 50 μm) executed together with binder-free rGO/CNT hybrids to form MSCs (rGO/CNT-MSCs, Figure 3.8) (Beidaghi and Wang 2012).

The implementation of CNTs within the graphene sheet layers inside planar interdigital microelectrodes improved the accessibility associated with the electrolyte ions present between the sheets of rGO and the power and energy densities. The MSCs based on rGO/CNT showed an areal capacitance corresponding to a value of ~6.1 mF cm^{-2} at 0.01 V s^{-1} and ~2.8 mF cm^{-2} with ~3.1 F cm^{-3} of stack capacitance at 50 V s^{-1}. All these values were considerably improved as compared to the values obtained for MSCs implementing pure rGO or CNTs (Z. S. Wu et al. 2012). The improved performance of the MSCs implemented with the rGO/CNTs is attributed to developed synergy leading to the heritance of the advantages of graphene, electrolyte-accessible, CNTs, and binder-free microelectrodes and the interdigital planar geometry in a single system. To attain large-energy density but keep satisfactory AC line filtering characteristics in a single device, Lin et al. reported 3D graphene/CNT carpet (G/CNTC)-based MSCs (G/CNTC-MSCs) with nickel substrate (Figure 3.9) (Z. S. Wu et al. 2010).

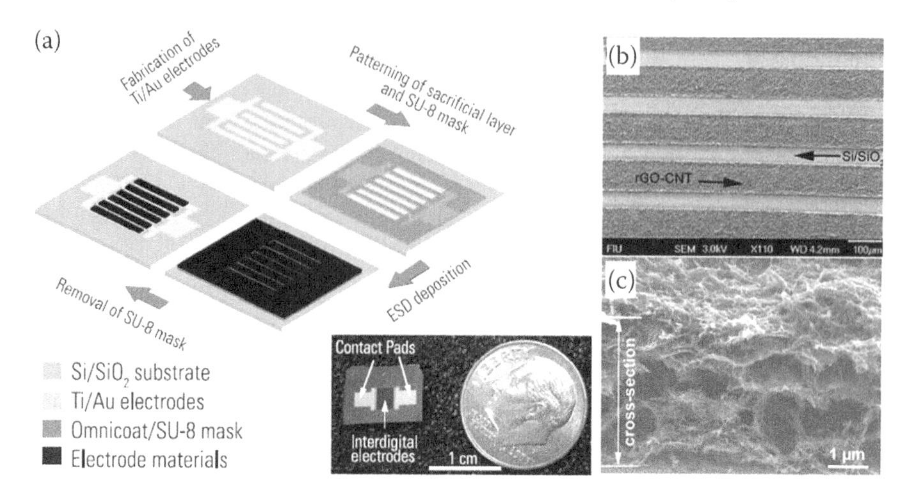

FIGURE 3.8 (a) Schematic of MSCs based on rGO/CNTs (inset: photograph of an MSC). SEM images (b) Top-view and (c) cross-section of interdigitated microelectrodes.

Source: Reprinted with permission from Beidaghi and Wang (Z. S. Wu et al. 2012). (Copyright 2012, John Wiley & Sons, Inc.)

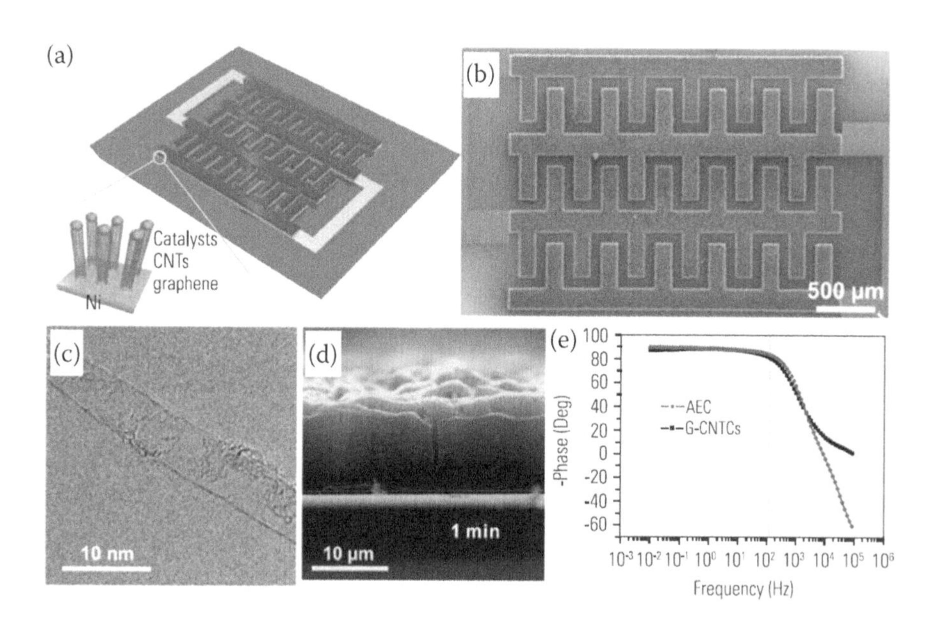

FIGURE 3.9 (a) The assembly of G/CNTC-MSCs. Inset: a pillar structure of Ni-G-CNTC. (b) SEM image of the MSCs. (c) TEM image of a SWCNT, (d) CNTCs. (e) Impedance phase angle vs frequency plot.

Source: Reprinted with permission from Lin et al. (Z. S. Wu et al. 2010) (Copyright 2013, American Chemical Society.)

The G/CNTC-MSCs showed an impedance phase angle of –81.5° at 120 Hz, which was very much like the commercial aluminum electrolytic capacitors (AECs, 83.9°) implemented for AC line filtering. When tested in the ionic liquid (BMIM-BF_4) G/CNTC-MSCs showed a high rate capability corresponding to a value of ~400 V s^{-1}, a volumetric energy density of 2.42 mWh cm^{-3} and a high power density of 115 W cm^{-3}. The excellent electrochemical performance showed by the G/CNTC-MSCs is attributed to unified graphene/nanotube junctions prevailing at the interface of the different carbon allotropes.

3.3.2 Graphene/Metal Oxide Hybrids

Peng and co-workers fabricated planar MSCs to enhance the charge-storage ability of the planar MSCs, implementing MnO_2/graphene sheets hybrid nanostructure (denoted MnO_2/GMSCs) and tested it in a gel electrolyte of PVA/H_3PO_4 (Figure 3.10) (L. Peng et al. 2013).

The flexible planar MnO_2/G-MSCs was prepared by implementing MnO_2/G hybrid film created with vacuum filtration concerning a solution of MnO_2/graphene. The next step was to transfer the thin film to a PET substrate, trailed by rasping the thin film to obtain slim strips to play the role of working electrodes. This step was followed by the Au current collectors' thermal evaporation on both faces of the substrate implemented with working electrodes, followed by covering with gel electrolyte on the parallel interspaces midway between the electrodes. This process led to the construction of a planar MnO_2/G-MSC (L. Peng et al. 2013). The hybrid thin graphene and MnO_2 sheets not only presented adequate

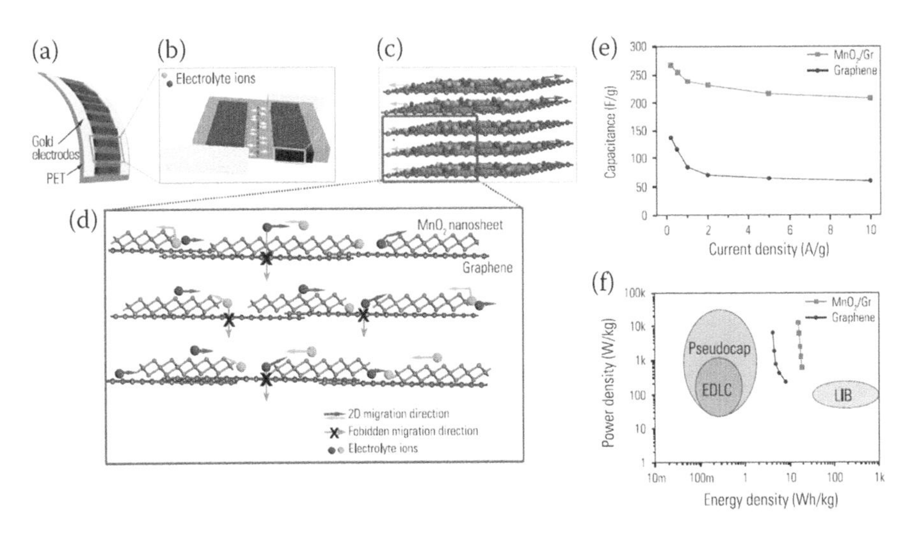

FIGURE 3.10 (a) Schematic diagram corresponding MSCs. (b) The planar MSC device. (c) The hybrid thin film with stacked geometry. (d) Schematic illustration of the ion transport. (e) Capacitance comparison of different MSCs. (f) A energy density vs power density plot of the MSCs.

Source: Reprinted with permission from Peng et al. (L. Peng et al. 2013). (Copyright 2013, American Chemical Society.)

electrochemically active surface leading to rapid desorption/absorption of electrolyte ions, but also availed extra interfacial area at the hybridized locations causing accelerated charge transport. The MnO_2/GMSCs procured showed augmented electrochemical charge-storage characteristics as compared to graphene-based MSCs, such as high specific capacitance, outstanding rate capability, and very high cyclic stability. Furthermore, the MnO_2/G-MSCs showed better flexibility and cyclic stability corresponding to initial capacitance retention greater than 90% following 1000 folding/unfolding cycles.

3.4 *IN SITU* AND *EX SITU* MATERIAL SYNTHESIS PARAMETERS

At the time of synthesis of any composite of two or more materials, *in situ* synthesis refers to the growth of any one material on the other material during combining the two materials. While *ex situ* synthesis involves the synthesis where the two parent materials are combined in the reaction. It has been seen in many cases that the structural and morphological properties vary completely for the composites synthesized *via in situ* or *ex situ*. The situation can be explained by the following example; according to a report, *in situ* electrospinning technique was employed to produce carbonized ZIF-67/carbon fiber electrodes as supercapacitor electrode material. To refrain from the inner pores of ZIF67 being clogged by the polymers and causing a reduction in the surface area and pore volume at the time of electrospinning process, an adapted slurry was incorporated during *in situ* electrospinning technique demonstrated in this report (Figure 3.11). The adapted slurry involves only PAN and cobalt salt. ZIF67 was created after integrating the cobalt/PAN electro spun matrix into ligand solution to refrain from PAN clogging the pores of ZIF67. It is important to mention that at the time of common straight electrospinning process, the slurry is made up of a metal salt, polymer, and a ligand precursor. A same process was demonstrated by Li et al., who prepared Co and N co-doped carbon for application in oxygen reduction reaction (B. Li et al. 2020). According to the literature, particle size ZIF67 was greater as compared to the diameter of nanofiber, leading to the gems-on-string structure having nanofiber crossing by the ZIF67. In this work, a bigger diameter value associated with the

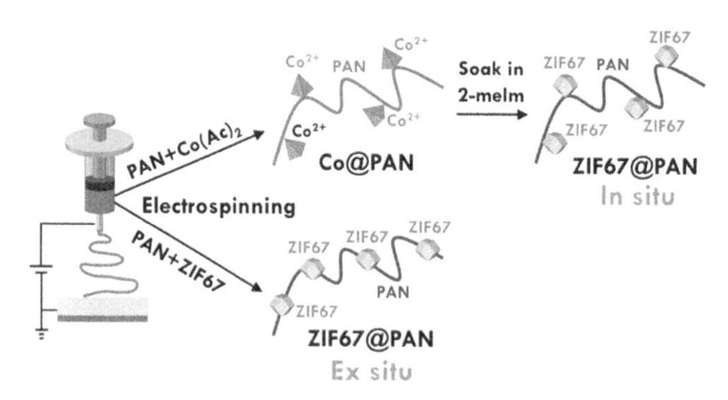

FIGURE 3.11 Illustration of the fabricating process for ZIF67@PAN Using *in situ* and *ex situ* methods. (C. H. Yang, Hsiao, and Lin 2021).

nanofiber was obtained and ZIF67 was excellently decorated on the inner as well as the outer surface of the nanofiber. Cobalt precursor concentrations were varied to fabricate ZIF67/ PAN electrodes, followed by the application of stabilization and carbonization methods to obtain carbonized ZIF67/carbon fiber (C67/PAN-OC) electrodes exhibiting large electrical conductivity. The enhanced *in situ* synthesized C_{67}/PAN-OC electrode exhibited the highest specific capacitance (C_F) corresponding to a value of 386.3 F g^{-1} at 20 mV s^{-1}. Whereas the C67/PAN-OC electrode material that was synthesized *ex situ* exhibited a CF value of 27.7 F g^{-1}. The augmented energy-storage capability associated with *in situ* synthesized electrode is attributed to the comprehensive growth of PAN-OC and the suitable carbonized ZIF67 contents, leading to a large electrochemical surface area (ECSA) and little resistances. A symmetric supercapacitor comprising of optimized C67/PAN-OC electrode material exhibits the highest energy density corresponding to a value of 9.64 kWh kg^{-1} at 0.55 kW kg^{-1} and a C_F retention of 59% following 1000 charge/ discharge cycles.

To have a better idea and explanation, Figure 3.12 demonstrates the developing method for *in situ* and *ex situ* synthesized C67/PAN-OC electrodes. The *ex situ* electrospinning process results in PAN to pass through ZIF67, resulting in destruction in the structure of ZIF67. During the process of carbonization, PAN gets transformed into N-doped carbon, and 2-melm in ZIF67 gets transformed to N and Co co-doped carbon. The various carbon materials formed from PAN and 2-melm exhibited different rising directions and compositions, which would abolish the original regular arrangement and reduce active surface area. At the last stage of carbonization, the carbon materials grown from 2-melm and PAN would grow concurrently and get integrated as a reliable carbon material, which can be seen by the one-body structure of the ex-situ synthesized C6714@PAN-OC electrode. Overall, the highest ECSA was attained by the incorporation of ZIF67 in PAN through *in situ* electrospinning approach 7 mmol L^{-1} of Co precursor concentration.

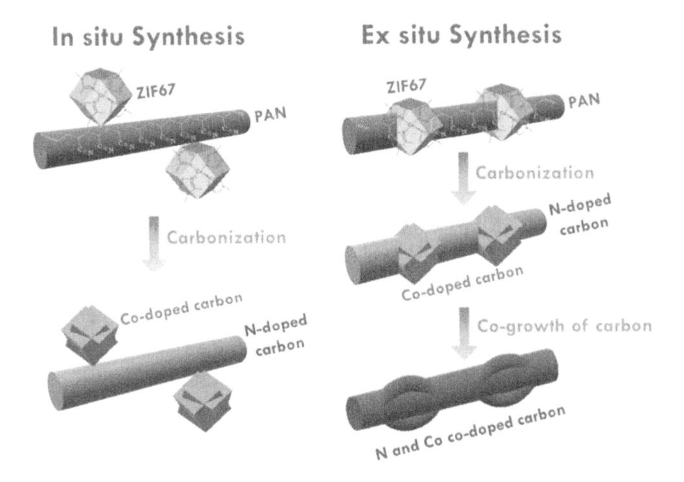

FIGURE 3.12 Illustration of *in situ* and *ex situ* electrospinning methods. (C. H. Yang, Hsiao, and Lin 2021).

It is observed that the electrospinning approach is more significant in rendering the ECSA as compared to the parameters used in the process of electrospinning such as concentration of the precursor. The specific surface areas 16.08 m^2 g^{-1} was obtained for the *in situ* and 22.92 m^2 g^{-1} was obtained for the *ex situ* synthesized C6714@ PAN-OC electrode material. Note that the trend followed in the specific surface area obtained through N_2 adsorption/desorption was inconsistent with the ECSA. In other words, the C6714@PAN-OC electrode synthesized *ex situ* exhibited little ECSA but a higher BET surface area as compared to the one synthesized *via in situ* approach. These results are attributed to the varying morphologies associated with the *in situ* and *ex situ* prepared C6714@PAN-OC electrodes. The *in situ* prepared C6714@PAN-OC electrode comprises of various particles visible on the surface. However, a much flatter surface was seen for the C6714@PANOC electrode synthesized by *ex situ* approach. It is concluded that the pores existing in the *ex situ* synthesized C6714@PAN-OC material are mostly distributed within the smooth layer. In other words, a thick layer was formed at the surface to limit the electrolyte from getting transported into the inner sides, those were occupied by several pores. Therefore, the acceptance of the active material by the electrolyte was much more easier in case of the *in situ* synthesized C6714@PAN-OC electrode than the electrode material synthesized *ex situ*. The N with lesser sizes are able to pass through the little holes created on the surface of the C6714@PAN-OC electrode synthesized *ex situ* and reach the inner pores and surfaces, but the electrolyte with higher size and larger viscosity as compared to the N is difficult to penetrate within the inner side of the *ex situ* synthesized C6714@ PAN-OC electrode *via* the little pores in its outer layer. This characteristics results in the larger surface area analyzed by the N_2 adsorption and desorption measurements in case of the *ex situ* synthesized C6714@PAN-OC electrode. The *in situ* formed C6714@PAN-OC electrode exhibits an average pore size and volume of 173.0 nm and 72.5 mm^3 g^{-1} respectively. The *ex situ* synthesized C6714@PAN-OC electrode, shows the average pore size and volume of 216.4 nm and 127.6 mm^3 g^{-1}. The bigger pore size and pore volume were seen for the C6714@PANOC electrode synthesized *ex situ*. Although the pore size is bigger in case of the *ex situ* synthesized C6714@PAN-OC electrode, the coat of the thick layer on the surface may restrict the diffusion of the electrolyte into the inner pores. However, implementing different approaches for determining the surface area of active materials may give opposite results. The ECSA is more consistent for approximating the amount of active sites of material as the ECSA measurement environment is similar to the energy-storage system. Influence of the concentration of cobalt precursor in an *in situ/ex situ* electrospinning processes on the electrochemical and physical performance of SC electrodes were investigated. Clear distinction of PAN substrate and ZIF67 could be obtained for *in situ* synthesized C67@PAN-OC electrodes. The *ex situ* synthesized C6714@ PAN-OC electrode exhibited a one-body structure created by the co-growth of carbon materials from the 2-mele and PAN. The *in situ* synthesized C6714@PAN-OC electrode exhibits the largest ECSA attributed to a suitable carbonized ZIF67 amount and whole growth of the PAN substrate; however the *ex-situ* synthesized C6714@PAN-OC electrode grants much lesser ECSA attributed to irregular growth and one-body nature of the carbon electrode. The largest C_F value corresponding to 386.3 F g^{-1} at 20 mV s^{-1} was achieved for the C6714@PAN-OC electrode synthesized *in situ*, but the *ex situ* synthesized C6714@PAN-OC electrode solitary shows a C_F value of 27.7 F g^{-1}. The minimum charge transfer resistance and series resistance values were obtained for *in situ* synthesized C6714@PAN-OC electrode. The supercapacitor containing *in situ* synthesized C6714@PAN-OC electrodes in combination with the gel electrolyte offers a potential window of 1.1 V, a highest energy density of 9.64 kWh kg^{-1} at 0.55 kW kg^{-1} and a Coulombic efficiency of 80% and a C_F retention of 59.5% following 1000 times charge/discharge cycles. This supercapacitor also exhibits a C_F retention of 75.6% as a result of bending at the angle of 90° for 100 times (C. H. Yang, Hsiao, and Lin 2021).

3.5 DEVICE DESIGN ARCHITECTURE

3.5.1 Device Configurations of Supercapacitor

The charge-storage capacity attainment of smart supercapacitors is mostly influenced by the configuration of the devices. However, generally, supercapacitors are fabricated in spiral wound or button cells, which become too bulky for smart energy-storage systems. Unlike the situation associated with the conventional supercapacitors, numerous microscale supercapacitor shapes have been proposed, according to the necessity of smart energy-storage devices. Growing numbers of smart electronic devices have conveniently increased the need for power sources, which will be flexible, thin, and foldable, such as electrochromic and shape-memory supercapacitors. Film supercapacitors are assembled as sandwiches in which all the components are intended to be very thin and often have excellent flexibility. However, conventionally used liquid electrolyte would face the problem of leakage at the time of bending or twisting, resulting in safety problem and damage. Hence, solid- or quasi-solid-state electrolytes are suitable for the design of these film supercapacitors. The electrolytes with solid- or semi solid-state form of very thin thickness will restrict the ion transport path length, along with serving as a sieve to further simplify the device configuration. A film electrochromic supercapacitor comprising solid-state with largely transparent $Ni(OH)_2$-polyethylenimine ethoxylated (PEIE)/poly(3,4-ethylenedioxythiophene): polystyrene sulfonate (PEDOT: PSS) polymer-based electrodes was reported (Ginting, Ovhal, and Kang 2018). Attributed to the electrode's thinness, the device showed high transparency and high coloration ability. Furthermore, diminishing the energy-storage device size within a chip will efficiently enhance the energy density of the devices. MSCs in general contain conductive substrates in the form of interdigitate as current collectors and various finger electrodes within the millimeter scale seized from each other by an insulating layer. The electrolyte is covered on the devices to restrict ion transport between electrodes. Smart micro-supercapacitors are possible through reasonabe design with the selection of suitable constituents. Furthermore, the micro-design in supercapacitors will abbreviate the diffusion path for ions in electrolyte, causing an effective utilization of the ECSA of micro-electrodes. The unbolted side edges of micro-electrodes remain in touch with the electrolyte, which paves the path for the electrolyte to infiltrate into the micro-electrodes, resulting in large energy density.

As compared to the other configurations corresponding to the supercapacitors, micro-supercapacitors are easy to integrate on the chip, and these are more companionable to be associated with the other micro-electronic devices, having the potential to act as smart power sources for the micro-electronic devices. As compared to the conventional fiber supercapacitors, planar supercapacitors remain 1D wires with diameters in the range from micrometer to millimeter. Three important configurations are related to the fiber supercapacitor devices: all-in-one fiber supercapacitors with twisted and coaxial fibers. The coaxial fiber supercapacitors are generally integrated by core fiber electrode, implementing layer-by-layer fabrication of electrode shell, and solid-state electrolyte or separator together. According to this configuration, a self-healing supercapacitor based on fiber was designed, where CNTs acted as working electrode and self-healing substance functioning as core. Furthermore, fiber architecture supercapacitors can be spun into textiles or fabrics, which exhibit significant prospects for working as wearable smart energy-storage devices. CNT fibers were warped beside the Ti wire to play the role of a fiber substance trailed by the deposition on the TiO_2 nanotube modified segments grown on the Ti wire to achieve photoelectric conversion. In case of the all-in-one fiber configuration, one fiber is deposited with the micro-electrodes followed by the covering of electrolyte. However, their implementation with the smart devices becomes difficult due to their complex fabrication procedure.

3.5.2 Self-Healing Supercapacitors

Flexible supercapacitors with various configurations have been designed and can very often tolerate bending or stretching strains to some extent (Niu, Dong, et al. 2013). However, these devices will be challenged by unwanted mechanical damage due to large external or internal deformation, causing significant deprivation of electrochemical performance and additionally causing severe safety concerns attributed to electrolyte leakage (X. Cheng et al. 2018). If supercapacitors exhibit the ability to repair on their own any damage, they will be able to regain the initial performance or minimize the distortion of the performance in association with the supercapacitors. Consequently, the lifetime and toughness of self-repairing supercapacitors will be enhanced to a large extent, minimizing the electronic cost and waste (Shuo Huang et al. 2019). In this regard, several materials possessing the ability to self-heal, particularly polymers, are implemented to fabricate and assemble self-repairing supercapacitors. These self-healing polymers mostly comprise significantly reversible active chemically and physically cross-linked bonds, which are capable of responding to external physical inducements, such as temperature, pH, and light, to achieve self-healing characteristics. The primary self-healing supercapacitor was designed in 2014 by implementing a self-healing substrate consisting of hierarchical flower-like TiO_2 nanospheres together with a supramolecular network (Figure 3.13A) (Hua Wang et al. 2014). As a result of the presence of a high amount of hydrogen bond donors and acceptors in the supramolecular network, when destroyed, the substrate was able to re-build the dynamical chain and the cross-links between the cracked surfaces. Moreover, in this situation, networks consisting of single-walled carbon nanotubes (SWCNTs) were distributed on

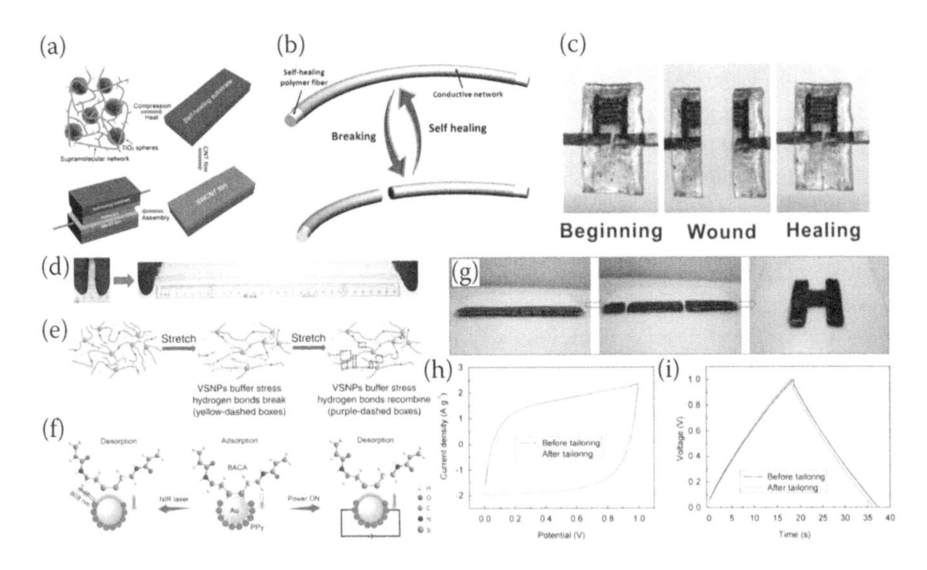

FIGURE 3.13 (a) Schematic illustrating the planar self-repairing supercapacitor. Reproduced with permission (Hua Wang et al. 2014). Copyright 2014, Wiley-VCH. (b) Self-repairing of fiber-like electrodes. Reproduced with permission. (H. Sun et al. 2014) Copyright 2014, Wiley-VCH. (c) Optical pictures of self-repairing micro-supercapacitor. Reproduced with permission. (Yue et al. 2018) Copyright 2018, American Chemical Society. D, Elongated and relaxed state of the self-repairing. E, Schematic illustrated the source of stretchability. Reproduced with permission. (Y. Huang, Zhong, et al. 2015) Copyright 2015, Nature Publishing Group. F, Illustration of the repairing method. Reproduced with permission. (C. R. Chen et al. 2019) Copyright 2019, Wiley-VCH. G, Digital images of couture all-repairable supercapacitor. H, CV curves. I, GCD curves.

Source: Reproduced with permission. (Z. Wang and Pan 2017) Copyright 2017, Wiley-VCH.

self-repairing substrate, which were used as both the electrodes of supercapacitors. Furthermore, based on the same self-repairing polymer, a polymer fiber showing healable characteristics was intended in 2014, where the aligned CNTs were assembled onto its surface; it was implemented as a self-healing supercapacitor electrode (Figure 3.13) (H. Sun et al. 2014).

Another self-repairing supercapacitor was retrieved exhibiting large mechanical strength from PU. The fiber-shaped supercapacitor exhibited 54.2% and 82.4% of capacity retention following third healing/cutting cycle and a 100% stretch, respectively. On the same note, a PU shell was integrated with 3D MXene-rGO composite to form micro-supercapacitors exhibiting self-repairing ability in 2018 (Figure 3.13C) (Yue et al. 2018), which exhibited a large areal capacitance of 34.6 mF cm^{-2} at 1 mV s^{-}. Gel or hydrogel-based electrolytes generally showed significant physical flexibility and outstanding electrochemical properties, where the main constituent is the polymer (Zhong et al. 2015). The conventional poly(vinyl pyrrolidone) (PVP) or poly(vinyl alcohol) (PVA) implemented gel electrolytes can easily self-heal due to the presence of O-HO hydrogen bonds amid the polymer chains (Shuo Huang et al. 2019). However, these gel electrolytes are formed in the nonattendance of cross-linking, leading to unacceptable mechanical and healable properties. To enhance the suitability of self-healing in addition to the gel electrolytes, a gel polymer electrolyte consisting of hybrid vinyl-silica nanoparticles (VSNPsPAA) and polyacrylic acid was designed successfully in 2015 (Figure 3.13 D, E) (Y. Huang, Zhong, et al. 2015). Credited to the undistinguished architecture of the cross-linked polyacrylic acid by vinyl hybrid silica nanoparticles possessing hydrogen bonds, that kind of electrolyte is largely stretchable and healable, even up to >3700%, resulting in ~100% retention of capacitance after 20 cutting/healing attempts. Unlike the cross-link in the polymer molecules within the hydrogel electrolytes, a hydrogel by Laponite cross-linked and graphene oxide was fabricated to be implemented as self-healing supercapacitors in 2019 (Huili Li et al. 2019). The functional groups, like COOH, OH, and Mg^{2+}, and GO in Laponite result in cross-linking of $CONH_2$ in the polymer chains, causing excellent mechanical stretchability and repairing performance. Because the synthesis of electrolyte in hydrogel form is a method of transferring liquid to hydrogel state, several components of a supercapacitor, including separator, cathode, and anode, is likely to amalgamate together through the hydrogel electrolyte and controlling the adjusting the supercapacitor configuration together with the gel process. The combined construction becomes very helpful in the method of seal-healing the entire devices. An all-healing real-time all-gel-state supercapacitor was designed in 2019 (Figure 3.13F) (C. R. Chen et al. 2019). The electrode was based on GCP@PPy hydrogel, and the CP hydrogel electrolyte possessed a chemical cross-linking with gold nanoparticles. The result exhibited a large stretching strain of 800% and rapid electrical repairability with 80% efficiency within two minutes. Similarly, an all-healable supercapacitor that could rebuild its electrochemical performances and configuration has also been reported in 2017 (Figure 3.13) (Z. Wang and Pan 2017).

3.5.3 Shape-Memory Supercapacitors

Supercapacitors can carry out diverse deformations irreversible in nature during practical application, which will cause functional and structural damage due to long-term stress (Xinyu Wang et al. 2017). Because of the discovery of the shape-memory function, the likely fatigue associated with the supercapacitors will meaningfully recover, expanding the cyclic life of the devices. Various kinds of shape-memory devices have been reported. To be compatible with the various shape-memory devices, the shape-memory supercapacitor is needed. Suitable shape-memory materials possess the ability to resort to external stimuli, like pressure, temperature, and magnetic force. Shape-memory supercapacitors are classified into two kinds: the first one is shape-memory polymers (SMPs), and the second one is shape-memory alloys (SMAs) (K. Guo et al. 2017). For SMA materials, the shape-memory effect is conducted by a reversible crystalline phase change, termed as martensiteaustenite transformation (Yun Yang et al. 2017). As a result of heating at a certain temperature, the distorted SMAs generally restore their initial shape, and all the plastic deformations are restored. SMAs are classified into four different classes, including Fe-based, NiTi-based, Cu-based, and intermetallic compounds (Z. G. Wei, Sandstroröm, and Miyazaki 1998). In all of these, NiTi forms the SMAs for supercapacitors credited to its larger mechanical and electrical characteristics. For example, a shape-memory supercapacitor was designed by implementing graphene coated on NiTi alloy flakes, with the negative electrode and very thin MnO_2/Ni film acting as the positive electrode in 2016 (Figure 3.14) (Lingyang Liu et al. 2016). In the ambient temperature, the destroyed device was able to recover its initial planar shape within 550 seconds. It could be shaped into a watchband, which would be able to automatically wrap itself onto the human wrist as it touches a human body. Nevertheless, the rate of recovery associated with this kind of shape-memory supercapacitor is lower in comparison, which can be attributed to the stress coming from the positive electrode. Fiber-shaped shape-memory supercapacitors show greater potential and design versatility in wearable electronic devices credited to their 1D structure and the capability of transformation to any shape (Zhibin Yang et al. 2013). In this case, in 2006, NiTi wires were applied as current collectors as a twisted wire-shaped supercapacitor. The shape-regaining method associated with this supercapacitor was accomplished within 25 seconds with capacitance retention of 96%.

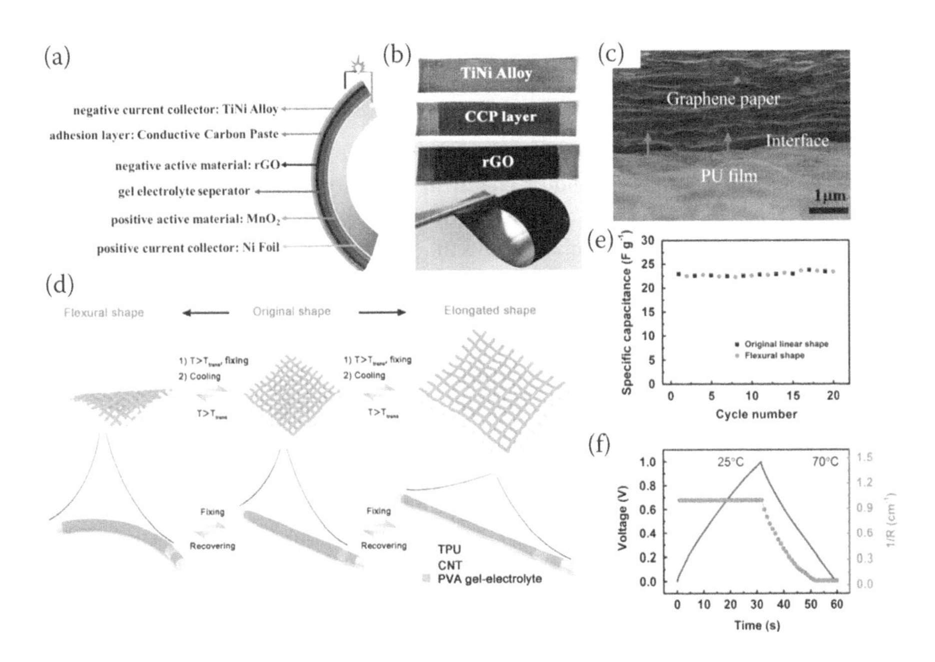

FIGURE 3.14 (a) Schematic of the shape-memory supercapacitor. (b) Optical images of the shape-memory supercapacitor. Reproduced with permission. (Lingyang Liu et al. 2016) Copyright 2016, Wiley-VCH. (c) Cross-sectional SEM image of the (PU) composite film. Reproduced with permission. (Tung et al. 2016) Copyright 2012, Elsevier. (d) Fiber-shaped shape memory supercapacitor schematic. (e) Variation in the specific capacitance with the cycle number. (f) Galvanostatic charge/discharge plots associated with the shape-memory supercapacitor.

Source: Reproduced with permission. (J. Deng et al. 2015) Copyright 2015, Wiley-VCH.

The supercapacitors were cast into traditional fabrics to design the smart sleeve, which was able to learn its shape and self-curl for understanding the heat exertion and cooling when the human body was too hot. SMPs including polyurethane, trans-isopolyprene, styrene-butadiene, and polynorbornene copolymers exhibited lighter weight as compared to the SMAs (J. Deng et al. 2015). Furtheromore, as a result of the combination of various reversible phase transitions in the polymers, SMPs are able to remember more than one shape and resume their original structure as subjected to different stimuli, such as light, magnetic fields, temperature, and electric currents. Therefore, SMPs are lately implemented in shape-memory supercapacitor devices. By coating graphene paper on the PU film, a double-layered composite film was designed; it exhibited outstanding shape-memory characteristics in 2012 (Figure 3.14) (Tung et al. 2016). The composite film possessed the capability to be transformed at 80 °C and ibe restored to its initial shape at ambient temperature in less than one second, exhibiting excellent shape-recovery capability. A shape-memory fiber-implemented supercapacitor was designed through sequentially coating thin layers of CNTs, PVA-gel electrolyte, CNTs and PVA gel electrolyte on to a PU fiber in 2015 (Figure 3.14 D-F) (J. Deng et al. 2015). Additional to its stretchability and flexibility, the supercapacitor was able to be "frozen" and transformed in various shapes and sizes opted by the user.

3.5.4 Electrochromic Supercapacitors

Electrochromic materials retain the ability to transform into various colors through insertion/extraction of charge and/or chemical reduction/oxidation. Electrochromic materials, if implemented as supercapacitor components, would result in supercapacitor storing energy in combination with accurate sensing of the variations in the energy showed by their detectable color change. The common electrochromic materials include conducting polymers, transition metal oxides, and metal organic frameworks, which can also be implemented as the electrode of the supercapacitors. Transition metal oxides were the pioneering material to be applied in electrochromic electronic devices. Among the various transition metal oxides, oxides of tungsten form the most general electrochromic materials, where the insertion/extraction of proton will lead to a change of color. The first work related to designing of electrodes for supercapacitors implementing WO_3 films on fluorine-doped tin oxide (FTO) glass wasattempted in 2014. The electrode material exhibited a high value of specific capacitance of 639.8 $F.g^{-1}$ in combination with excellent electrochromic properties (Figure 3.15) (P. Yang and Mai 2015). The color transformation went from transparent to deep blue, accompanied by an abrupt minimization in the optical transmittance from 91.3% to 15.1% at 633 nm from 0 to –0.6 V (vs Ag/AgCl).

The electrochromic system that will bring a change in the value of transmittance in response to incident sunlight can be implemented in wearable electronics and architecture; the devices will be capable of reversibly transforming their color reliably and aptly store energy. In 2018, a nano-generator in combination with an electrochromic micro-supercapacitor was reported. In the device AgNWs/NiO on ITO electrodes were implemented as the capacitive material, as well as the

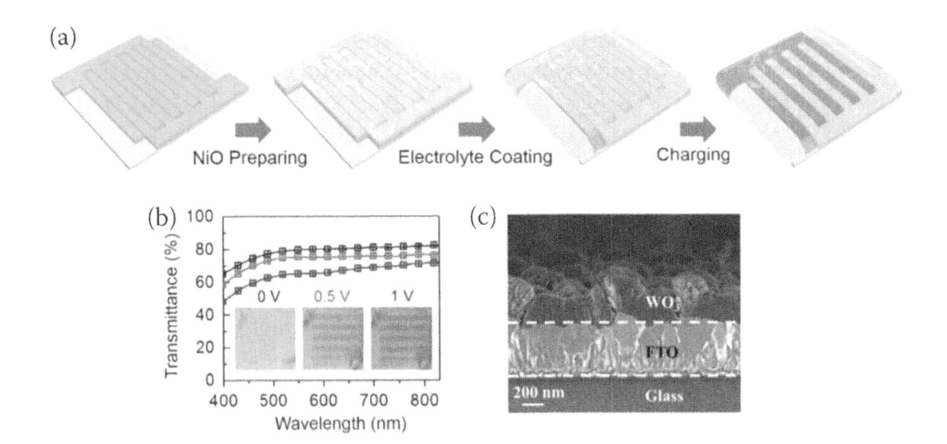

FIGURE 3.15 (a) Schematic of the electrochromic micro-supercapacitor. (b) Digital pictures of electrochromic micro-supercapacitor and corresponding transmittance plotted with the wavelength. Reproduced with permission. 66 Copyright 2018, Wiley-VCH. (c) Cross-sectional scanning electron microscopy image of the WO_3 thin film.

Source: Reproduced with permission. (P. Yang and Mai 2015) Copyright 2015, Wiley-VCH.

electrochromic material (S. Qin et al. 2018). During the charging/discharging method, the positive electrodes of the micro-supercapacitor changed to dark color and got back to its original transparency due to reversible faradaic redox process of Ni^{2+}/Ni^{3+} couple (Rui Wang, Yao, and Niu 2020).

3.6 NEW DEVICES AND APPLICATIONS OF SUPERCAPACITOR

3.6.1 Battery-SC Hybrid (BSH) Device

Battery-supercapacitor hybrid (BSH) devices consider themselves significant due to their projected utilization in different smart optoelectronic devices, electrical networks, and electric vehicles, etc., with traditional Ni-MH, P-acid, Ni-Cd, Li-ion batteries (LIBs) in combination with some advanced batteries, including Li-sulfur, Li-air, Al-ion, and Na-ion other metal-ion-based aqueous batteries. Energy-storage device fabrication through the combination of an electrode with higher value of (charge storage ability) C_s in combination with a greater (power delivery) P_s electrode with a high capacity, termed as battery-supercap hybrid (BSH), projects a vital method for achieving a system exhibiting the characteristics of both battery and supercapacitors (F. Zhang et al. 2013).

In Li-ion BSH, the supercapacitor-based electrode materials are mainly carbon-based materials, including graphene, activated carbon, Carbon Nano Tube, etc., and the battery material consists of metal oxides, intercalation compounds, and their composites. Zhang et al. (F. Zhang et al. 2013) fabricated the EDLC type (positive) electrode and battery type (negative) electrode and amalgamated them to design a device, which was able to deliver excellent electrochemical performance corresponding to energy density of 147 Wh kg^{-1}

at power density of 150 W kg^{-1} in combination with good capacitance retention. A novel supercapacitor/Li-ion battery (SC/BT) technique coined as hybrid energy storage system (HESS) was attempted by Peng et al. (Lx et al. 2017) to be applied in electric vehicles (EV) through an ADVISOR simulator. In this system, the brake-regeneration energy is harnessed from the supercapacitor package. Limited availability of Li resources has proved to be a challenge to the researchers to rely on other abundantly available earth metals. Na-ion supercapacitors are one such category of devices being investigated, in combination with high-capacity battery electrode and high-power density-based SC electrode to result in an energy-storage system having high rate capacity. Lu et al. (K. Lu et al. 2015) designed a suave Na-ion supercapacitor electrode implementing Mn hexacyanoferrate acting as the cathode and Fe_3O_4/rGO acting as the anode. The device was tested in aqueous electrolyte with working potential of 1.8 V, delivering a energy density of 27.9 W h kg^{-1} and a power density of 2183.5 W kg^{-1} in combination with 82.2% capacitance retention following 1000 cycles.

3.6.2 Electrochemical Flow Capacitor (EFC)

In electrochemical flow capacitor (EFC), energy storage occurs in double layers comprising the charge of carbon molecules. In this type of device, a slurry mixture comprising carbon electrolyte plays the role of an active material responsible for the storage of charge. EFC comprises a cell casing the external deposits originating from a reservoir consisting of a mixture of carbon materials and electrolytes. The uncharged slurry is transported from the stock tanks in the flow cell energizing the carbon material. Following the loading of the charge, the slurry can be taken in huge tanks until the time the energy demand increases; as the need increases, there is setback of the entire process. EFCs can sustain large number of load (charge-discharge) cycles (Presser et al. 2012). A liquid electrode consisting of HQ/carbon spheres resulting in a greater capacitance reaching a value of 64 F g^{-1}, which is 50% larger as compared to the flowable type electrodes based on carbon material (Y. Guo et al. 2021).

3.6.3 Alternating Current (AC) Line-Filtering Supercapacitors

A supercapacitor is apt to replace bulky Al-based electrolytic capacitors (AEC), broadly been applied in the AC line filter, leading to miniaturization of the device. However, the supercapacitors implemented for this cause exhibit a voltage window limited to ~20 V. To augment the voltage window, fabrication of appropriate carbon electrodes possessing a suitable pore structure is significant. Yoo et al. (Y. Yoo et al. 2016) attempted a mesoporous carbon that was graphene based as a supercapacitor electrode, and the corresponding supercapacitor exhibited a 2.5 V of voltage window with a surface capacitance corresponding to 560 mF cm^{-2} in combination with rapid frequency response (ϕ–80°) produced at 120 Hz. Integration of a small amount CNT leads to enhanced voltage of 40 V. Rangom et al. (Rangom, Tang, and Nazar 2015) attempted the fabrication of self-supporting electrodes comprising of single-walled carbon nano tubes (SWCNTs). Mesoporous 3D electrodes based on SWCNTs result in ion transfer in thick films and lead to improvement in 120 Hz ac line frequency. Wu et al. (Zhenkun Wu et al. 2015) reported the implementation of high-scale graphene ac line filters (Qi et al. 2017). In the particular work, the reduced GO (rGO) possessed metal joints applied as the electrode material and the fabricated device was tested at the phase angle of −75.4° at 120 Hz, having a time constant of 0.35 ms, an areal capacitance of 316 μF cm^{-2} in combination with 97.2% of C_s over 10,000 cycles. Kurra et al. (Kurra, Hota, and Alshareef 2015) studied the PEDOT based micro-supercapacitor operated in 1 M H_2SO_4 at the scan rate of 500 V s^{-1} at the frequencies of 400 Hz at ~−45 ° and obtained 9 mF cm^{-2} of areal capacity. The capacitance retention exhibited was 80% following 10,000 cycles, in combination with a Coulombic efficiencies (η) of 100% and energy density of 7.7 mW h cm^{-3}.

3.6.4 Thermally Chargeable Supercapacitors

Recent research shows that the thermal energy earlier projected to have less contribution can be harnessed for power supplies in portable electronics. Energy conversion obtained through thermoelectric is an apt approach to channelize waste heat. Nevertheless, the disadvantages include small output working potential, in combination with incapacity to stock energy requiring other components. Supercapacitor with the ability of thermal self-charging comprises of the Seebeck effect, temperature-influencing thermally activated ion diffusion and electrochemical redox potential. It includes two electrodes kept at two separate temperatures (Härtel et al. 2015). A new approach to produce large operating potential in combination with a gradient of temperature, including traditional thermoelectrics, has been attempted (S. L. Kim, Lin, and Yu 2016). According to the report, PANI-coated CNT and graphene electrodes arrest polystyrene sulfonic acid (PSSH) films, in which the electrochemical reactions excited thermally leads to recharging the supercapacitor without the requirement for an external energy source. In the presence of a small temperature gradient of 5 K, the supercapacitor thermal charger generates 38 mV of voltage and charge-storage capacity of 1200 F m^{-2}. Al-zubaidi et al. (Yaseen et al. 2021) stated the incidence of thermal induction at the interface of the solid-liquid and the ionic electrolyte. Further, the research also showed the self-charging in the supercapacitors at the moment of thermal excitation. Wang et al. (Jianjian Wang et al. 2015) fabricated supercapacitor electrode material by applying phenomenon of thermal charge to retrieve the wasted energy in supercapacitors following the method of charging/discharging. Zhao et al. (D. Zhao et al. 2016) implemented an electrolyte consisting of an asymmetric polymer retrieved from NaOH-treated polyethylene oxide (PEO-NaOH) to produce stress by implementing thermal induction in supercapacitor. The electrodes implemented were of Au and multi-walled carbon nanotubes (MWCNTs) reinforced on Au and could attain a thermal voltage of 10 mV K^{-1}, charge capacity (area) of 1.03 mF cm^{-2} and energy density of 1.35 mJ cm^{-2} at the temperature gradient of 4.5 K.

3.6.5 Piezoelectric Supercapacitors

Improving the limit of integration, in combination with minimizing the loss of energy in circuits of power management, becomes a mandatory component in the current situation. Generally, a full-wave rectifier is implemented in combination with the energy-storage device, along with the nanogenerator based on piezoelectric component, which is expected to diminish the density of integration and enhance the energy loss. Recently, Xing et al. (Xing et al. 2014) attempted to fabricate the self-charged supercapacitor by integrating piezoelectric separator into the Li-ion battery. However, attributed to sluggish charge and abridged span of the Li-ion battery, supercapacitors have garnered tremendous attention. Song et al. (R. Song et al. 2015) implemented a PVDF film into supercapacitor as both separator and current collector. A supercapacitor was fabricated with the help of PVDF film acting as an anode and a carbon cloth in combination with H_2SO_4/PVA electrolyte acting as the cathode. The PVDF piezoelectric film exhibited a charge storage of 357.6 F m^{-2}, an energy density of 400 mW m^{-2} and a power density of 49.67 mW h m^{-2}. Relating to the capability to lead to high performance, the choice is presented also based on their physical properties including pressure, temperature, stress, self-healing property, battery type, size, AC current type, etc. Ramadoss et al. (Ramadoss et al. 2015) attempted to study a new material synthesized by integrating the piezoelectric type and the pseudocapacitors for the storage of energy.

4 Supercapacitor Materials

The supercapacitor's performance depends on various parameters that depend on the material's properties. Hence, researchers are devoted to altering and/or optimizing these properties by designing novel electrode materials either by combining two or more materials within a single system, doping, or surface functionalization, etc. Some of the properties and their effect in the capacitive performance are listed below:

Crystallinity: It is well known that metal-based compounds like metal oxides, metal nitrides, and metal sulfides generally are crystalline. An increased level of crystallinity makes a material more rigid and lessens the material's charge-storage capacity. For example, with an increase in the value of crystallinity, the specific capacitance was observed to be reduced as reported by Kim and Popov (H. Kim and Popov 2003) and shown in Figure 4.1 (a) and (b). The crystallinity of the material increases with increased annealing temperature, evident by XRD pattern; it has the opposite effect on the material's capacitance. This effect could be attributed to the limited transport of the charges species/electrolyte to the bulk of the crystalline material (Merlet et al. 2012). The redox materials' increased crystallinity inhibits the bulk diffusion within the electrolyte of the material and limits the specific capacitance contribution, but only from the surface-redox reactions (Y. Liu, Jiang, and Shao 2020). On the other hand, the amorphous redox materials' specific capacitance can be contributed from both the exterior of the electrode as well as bulk interaction between the electrode and the electrolyte.

Porosity: Specific capacitance is very significantly dependent upon the pore features of any electrode material. Pore features also determine the active surface area and the pore volume of an electrode material (Heimböckel, Hoffmann, and Fröba 2019). The electrolyte diffusion and internal conductivity depends on the pore structures and pore size distribution of the electrode materials. The specific surface area of the electrode material can be determined through adsorption-desorption experiments by employing the Brunauer–Emmett–Teller (BET) analysis of the obtained isotherms. Further, the Barrett-Joyner-Halenda (BJH) model can be employed to get particle-size dispersal, as well. The influence of pore-size distribution on the capacitance of the material has an intriguing effect, and it depends on the thickness of the double layer, as shown in Figure 4.2 (Heimböckel, Hoffmann, and Fröba 2019). Electrode materials can be categorized as microporous (<2 nm), mesoporous (2–50 nm), and macroporous (>50–200). Micropores promote the increase in surface area but do not promote the electrolytic ion diffusion. On the other hand, macropores are not favorable to causing a high surface area, but they favor ion diffusion. Mesopores can help to promote both ion diffusion and surface area, but with compromising both to some extent. Hence, to get the best of the pore features, materials' hierarchical pore structures are gaining researchers' attention.

DOI: 10.1201/9781003174554-4

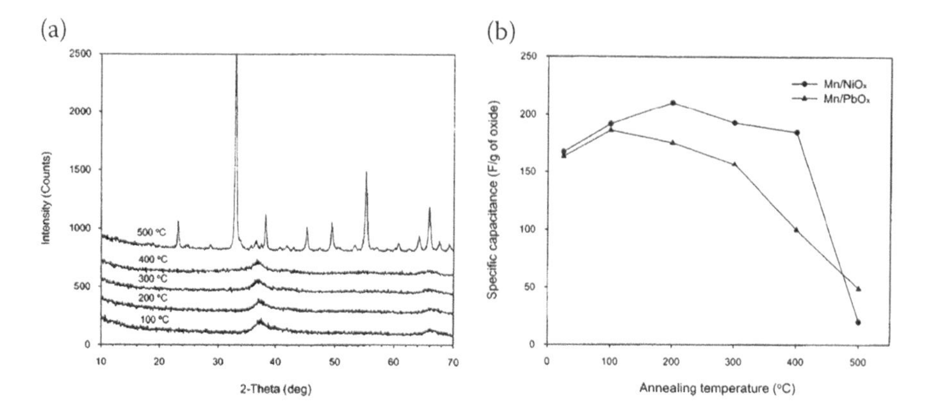

FIGURE 4.1 (a) XRD pattern as function of annealing temperature from Mn/NiO_x and (b) Effect of annealing temperature/crystallinity on the specific capacitance of Mn/PbO_x and Mn/NiO_x.

Source: Reprinted with permission from (H. Kim and Popov 2003). Copyright ©2003, IOP Publishing, Ltd.

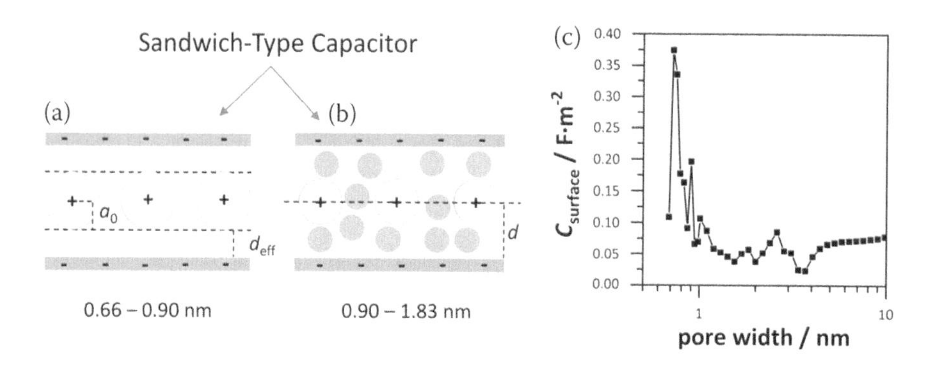

FIGURE 4.2 (a) and (b) The sandwich double-layer capacitor with different pore sizes showing average ion radius a_o and variation of double-layer thickness d_{eff} (c) Surface area normalized capacitance at different pore size.

Source: Reprinted with permission from (Heimböckel, Hoffmann, and Fröba 2019), Copyright © 2019, Royal Society of Chemistry.

Wettability: This property is another significant aspect that determined the level of the favorable association (physical or chemical) of the electrolyte with the electrode active material. The reversible redox reaction in case of the redox active material crucially depends on the exchange of the charges species and the electron hopping. The hopping of electrons actually takes place between the OH^- and the H_2O sites of the electrode material. Hence, physically, a chemically bound water to the active electrode material surface plays a major role in determining the interaction limit of the electrolyte and the charge propagation.

Conductivity: The intrinsic conductivity of an electrode material plays a crucial role in the the electrode material's resulting total resistance. Doping of the electrode material is often performed to result in enhanced conductivity. Better conductivity leads to easy-charge transport within the electrode material, resulting in higher power density, better reversibility, and higher specific capacitance.

Size of the material: Nanosized particles are known to show better electrochemical charge-storage properties as compared to the bulk materials. Hence, nanostructured materials gained a lot of attention due to their promising charge-storage aspect. Nanostructured materials always provide a large number of active sites for electrode-electrolyte interactions. Hence, in general, the smaller-sized materials exhibit augmented gravimetric capacitance. The size of the electrode material can effectively be adjusted by controlling the synthesis path and selecting appropriate precursors.

4.1 POROUS MATERIALS

4.1.1 Carbon Materials

This type of material in various forms is the most popularly used electrode material for fabricating supercapacitors attributed to high surface area, low cost, abundance, and well-known production methodology of the electrode. Carbon materials store charge electrostatically by the formation of an electric double layer at the electrode/electrolyte interface. Hence, for the carbon materials, the capacitance mainly originates from the surface area the electrolyte ions access. Significant factors controlling the electrochemical characteristics of the carbon-based material are their specific surface area, pore texture and shape, surface functionality, pore size distribution, and electrical conductivity (H. Yang et al. 2017; A. P. Singh et al. 2015). High surface area open to the electrolyte leads to high capability of storing a large amount of charge in the separated charged double layer at the interface of the electrode and the electrolyte. Besides the high surface area and porosity, the surface functionalization may also play an important role in enhancing the capacitance of the material. Certain forms of carbon, which are popularly studied and used as supercapacitor electrode material, are carbon aerogels, activated carbon, carbon nanotubes, graphene, carbon nano-onions, etc.

4.1.2 Activated Carbon (AC)

AC is the widely applied carbon form for supercapacitor electrode attributed to its good electrical conductivity, large surface area, and cost effectiveness. AC can be synthesized either by chemical or physical activation of several kinds of carbonaceous materials (e.g., coal, carbon containing biowaste, wood, etc.). The physical activation takes place by the carbon precursors' treatment at a large temperature (700–1200°C) in the presence of oxidizing gases like CO_2, steam, and air. Unlike the physical activation, the chemical activation is performed at a lesser temperature (400–700°C); implementing activating agents include potassium hydroxide, sodium hydroxide, phosphoric acid, nitric acid, and zinc chloride (Pandolfo and Hollenkamp 2006). Zeng et al. showed the effect of nitric acid activation of carbon,

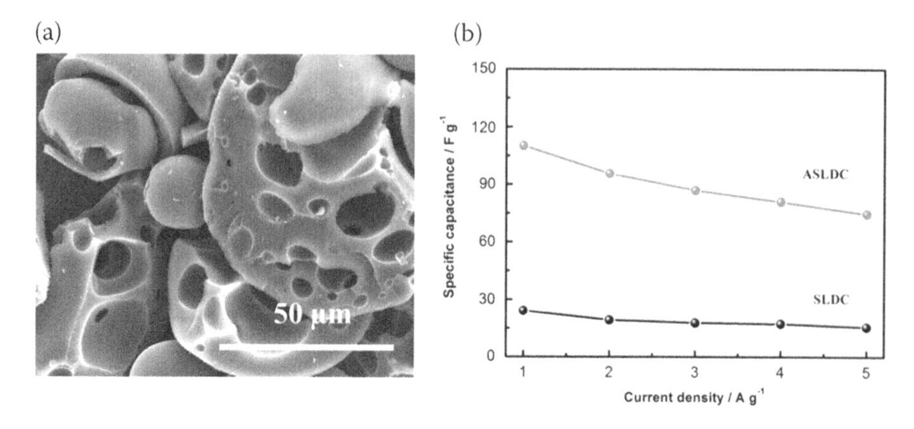

FIGURE 4.3 (a) SEM image of activated carbon derived from activated carbonaceous mudstone and lignin-derived carbons (ASLDC) and (b) Effect of activation on the specific capacitance of the material.

obtained from activated carbonaceous mudstone and lignin-derived carbons (ASLDC), on specific capacitance, as shown in Figure 4.3 (L. Zeng et al. 2019).

On the basis of the carbon precursors and triggering methods implemented, activated carbonexhibitsseveral physiochemical properties with high surface areas of till 3000 m^2 g^{-1}. Porous-activated carbon synthesized by applying the chemical or physical activation leads to a broad pore size distribution comprising micropores (<50 nm) (Saleem et al. 2019). Numerous researchers have assessed the influence of the pore structure and specific surface area (SSA) and the specific capacitance of the porous carbon. According to a report, AC with a large SSA value near to 3000 m^2 g^{-1} exhibited not very satisfactory specific capacitance value, implying an ineffectiveness of a large portion of the pores in the charge storage. Hence, for a satisfying performance, apart from the SSA, some other features also play a crucial role influencing the electrochemical performance to a large extent, including the pore size distribution (Lota et al. 2013; Lufrano and Staiti 2010). In addition, unnecessary activation causes in high pore volume, in turn causing setbacks including less conductivity and less density of material. These will in turn result in a loss of energy density and power capability. Measures have been put to study the impact of various electrolytes on the charge-storage performance of the activated carbon. It was observed that the charge-storage capacity of AC is larger in aqueous electrolytes (100 F g^{-1} – 300 F g^{-1}) than in the organic electrolytes (<150 F g^{-1}) (Hasan and Lee 2014).

4.1.3 Carbon Nanotubes

The discovery of CNTs ushered the path of advancement in the engineering as well as science of carbon-based materials. The overall resistance of all the components of a supercapacitor determines its power density. A significant amount of attention has been dedicated to CNT as supercapacitor electrode material attributed to its unique pore structure, improved mechanical and thermal stability, along with

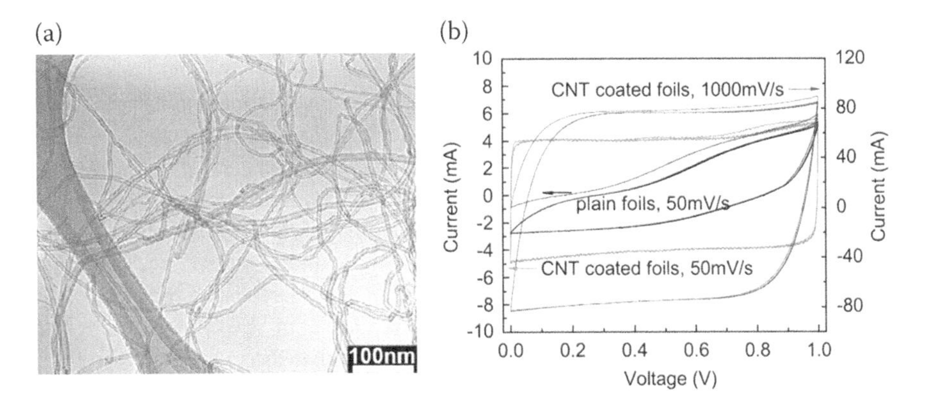

FIGURE 4.4 (a) TEM image of CNTs and (b) CV curves of plain Ni foil as current collector and CNT-coated Ni foils showing enhancement in current response in presence of CNT for the same scan rate.

Source: Reprinted with permission from (C. Du and Pan 2006), Copyright © 2006 IOP Publishing.

superior electrical properties (Q. Cheng et al. 2011a; Tamilarasan, Mishra, and Ramaprabhu 2011; C. Du and Pan 2007) CNTs are synthesized by the means of catalytic decomposition of hydrocarbons and precise manipulation of various parameters; it becomes probable to produce nanostructures in various shapes and also regulate their crystalline structure (Pandolfo and Hollenkamp 2006). A specific property of carbon nanotube that makes it different from the other carbon-based electrodes is that in CNTs the mesopores are interconnected allowing a perpetual distribution of charge, which brings almost entire accessible surface area to interact with the electrolyte and enhances the electrochemical response, as shown in Figure 4.4. CNTs have a lesser electrochemical series resistance as compared to the activated carbon attributed to the diffusion of the electrolytic ions into the mesoporous network.

CNT can be classified as MWCNTs and SWCNTs, or both are mostly studied as electrodes for supercapacitors. The high surface area reachable to the electrolyte and high conductivity renders CNTs to exhibit high specific power. Additionally, CNTs can act as efficient scaffold for active materials attributed to their open tubular network and large mechanical resilience. In general, CNT possess less SSA ($<500\ m^2/g$), further leading to reduced energy density than the energy density offered by the AC. CNTs can be activated chemically with the help of KOH, leading to augmentation in its specific capacitance. The above procedure can result in significant augmentation in the surface area of CNT (two to three folds), while maintaining its nanotubular morphology (Pandolfo and Hollenkamp 2006).

4.1.4 Graphene

Graphene has garnered attention as 2D-layered material for various applications, including energy storage, photonics, sensing, water splitting, drug delivery, etc. (Mathew and Balachandran 2021). Graphene exhibits a one-atom thick 2D-layered structure that has arisen as a distinctive carbon material having huge potential as the electrode material attributed to its excellent chemical stability, large conductivity, and high surface area (Shrestha et al. 2020). Recently, it was projected that graphene is applicable as material for supercapacitors, and the capacitance for this material in independent of the distribution of pores at solid state, unlike the other carbon materials, including CNTs, AC, etc. (Karthika, Rajalakshmi, and Dhathathreyan 2012; T. Kim et al. 2013).

Among the various carbon materials applied for the electrode materials in the electrochemical double-layer capacitors, graphene exhibits larger SSA near to 2630 $m^2 g^{-1}$ (T. Kim et al. 2013). If the entire SSA will interact with the electrolyte, graphene can exhibit a capacitance of 550 F g^{-1} (C. Liu et al. 2010). Another benefit of implementing graphene as an electrode material is that the main graphene sheet surfaces are exterior, making them readily available to the ions. Several distinctive processes presently are being investigated for designing different kinds of graphene, including micromechanical exfoliation, CVD, unzipping of CNTs, the arch-discharge method, electrochemical and chemical methods, epitaxial growth, and intercalation methods in graphite (S. M. Chen et al. 2014). To use the largest intrinsic specific surface area and surface capacitance associated with the single layer graphene, steps were taken to limit restacking of the graphene sheets during the synthesis of graphene and successive procedures for electrode production. These can be guaranteed by making arched graphene sheets that will refrain from restacking face to face. Energy density equal to 85.6 Wh kg^{-1} at RT and 136 Wh kg^{-1} at 80°C obtained at a current density of 1 A g^{-1} were obtained. The energy densities delivered by this material could be compared to that of battery implemented with Ni metal hydride (C. Liu et al. 2010). There are several processes of producing graphene from graphite. Chemically modified graphene (CMG) was obtained and tested implementing both aqueous and organic electrolyte. Capacitances of 99 and 135 F g^{-1}, respectively, were delivered by it (Aricò et al. 2010). Graphene is challenged by agglomeration of the layers and tends back to the formation of graphite. Graphene material implemented in supercapacitor electrodes can result in a large capacitance of 205 F g^{-1} at the 1 V when measured in aqueous electrolyte, associated with 28.5 Wh kg^{-1} energy density. These values are greater than the supercapacitors assembled using carbon-based electrode material. The experiment was carried out on a single-layered graphene-oxide sheet, which was reduced by hydrazine at room temperature (RT). Graphene synthesized using this procedure exhibited little less aggregation as compared to the chemically modified graphene synthesized by implementing aqueous solution (Pope et al. 2013). To establish a more effective technique for implementing graphene as an electrode in supercapacitor, three distinct processes were investigated. The first process involved thermal exfoliation of graphitic oxide, the second method involved heating nanodiamond at 1650°C in a He atmosphere, and the third and last method implemented the decomposition of camphor over nickel nano-particles to obtain graphene. The largest specific capacitance of 117 F g^{-1} associated with an energy density of 31.9 Wh kg^{-1} were exhibited by the graphene obtained by the thermal exfoliation of graphitic oxide (Iro, Subramani, and Dash 2016). For high-power applications of supercapacitor, it's important to design supercapacitors with large specific capacitance and rapid charging time obtained at a high current density.

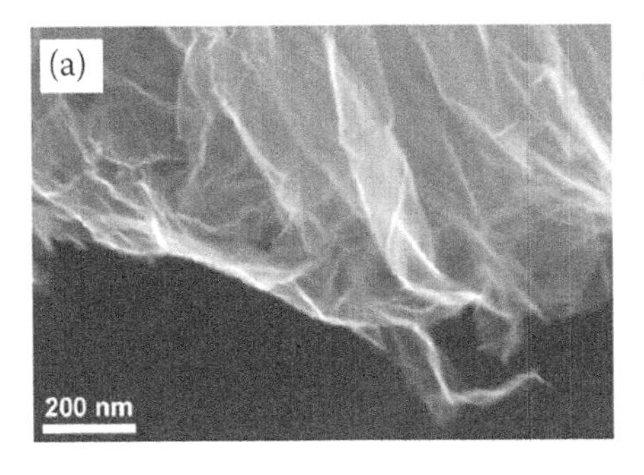

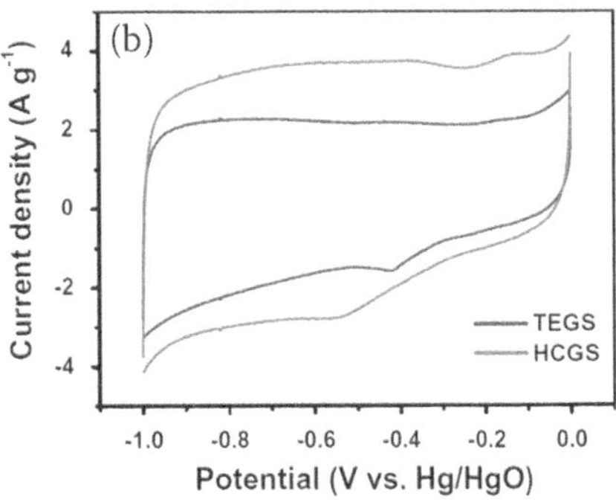

FIGURE 4.5 (a) SEM image of highly corrugated graphene sheet (HCGS) and (b) comparison of CV curves of thermally expanded graphene sheet (TEGS) and HCGS under 20 mV/s scan rate and 6 M KOH aqueous electrolyte solution.

Source: Reproduced with permission from (J. Yan et al. 2012), Copyright © 2012, Elsevier.

Graphene synthesized by modified hummer's method and tip sonication could exhibit such high value of specific capacitance and high power. Various charging current were investigated, including 2.5 A g^{-1}, 5 A g^{-1} to 7.5 A g^{-1}. Even at a large current of 7.5 A g^{-1} the energy and power density exhibited by the graphene was 58.25 Wh kg^{-1} & 13.12 kW kg^{-1}, respectively. This performance was adequate for their application in electric vehicle (Kakaei, Esrafili, and Ehsani 2019). Exfoliation of graphene is performed at a high temperature. A new method was implemented and investigated in which low temperature was applied for the exfoliation of graphene; this ability was attributed to its unique surface chemistry, which was shown by graphene exfoliated at low temperature. It also exhibited good energy-storage properties, and the capacitance exhibited was larger than the graphene exfoliated at high temperature (Lv et al. 2009). Several methods are being investigated on the possible way to reduce the problem of restacking and agglomerating graphene. Highly ridged graphene sheets were procured by thermally reducing the graphite oxide at a large temperature and then cooling it rapidly by implementing liquid nitrogen. A high capacitance of 349 F g^{-1} was exhibited by the material by highly corrugated graphene sheet (HCGS) and shows improved electrochemical performance in comparison with thermally expanded graphene sheet (TEGS) shown in Figure 4.5 (J. Yan et al. 2012).

4.2 METAL OXIDES

Metal oxides can be considered as another substitute for the materials implemented in electrode fabrication in supercapacitor since they exhibit large specific capacitance and small resistance, leading to simpler supercapacitor construction associated with large energy and power. The popularly implemented metal oxides as supercapacitor electrodes are ruthenium dioxide (RuO_2), nickel oxide (NiO), iridium oxide (IrO_2), manganese oxide (MnO_2), etc. The expense synthesized metal and its compatibility with a milder electrolyte make these metal oxides a possible attractive substitute (Shafey 2020).

4.2.1 Ruthenium Oxide (RuO_2)

RuO_2 in both amorphous and crystalline forms is significant for theoretical as well as practical purposes, attributed to distinctive combination of properties, including metallic conductivity, catalytic activities, electrochemical reduction-oxidation characteristics, large thermal and chemical stability,. and property of field emission. Equipped with these properties, RuO_2 becomes apt for various products involving resistors, ferroelectric films, electronics, and integrated circuit designing. Additionally, recent advances in the fabrication techniques and demand of flexible electronics led to RuO_2 flakes growing over flexible Kapton tape via precursor coating followed by Laser scribing, as shown in Figure 4.6 (K. Brousse et al. 2018). RuO_2 is also well-known supercapacitor electrode material storing charge through redox interaction with the electrolyte (Cuimei Zhao and Zheng 2015). Among the several metal oxides, those are implemented as the electrode materials are RuO_x, NiO_x and IrO_x, etc. RuO_2 is the most successful metal oxide attributed to various advantages, including long cycle life, broad potential window, large specific capacitance, largely reversible oxidation-reduction reaction, along with metallic conductivity. For application as a supercapacitor electrode, RuO_2 was produced electrochemically by implementing an electrodeposition approach. The obtained electrodes showed excellent cyclic stability delivering specific capacitance of 498 F g^{-1} (Gujar et al. 2007).

4.2.2 Nickel Oxide

Nickel oxide is one of the very promising electrode materials for application as electrode material for supercapacitor attributed to its environmental friendliness, facile synthesis, and cost effectiveness. Among the various advantages for electrochemical strategy are simplicity, reliability, accuracy, cost effectiveness, and versatility. Using an electrochemical approach, nickel hydroxide can be changed into nickel oxide. The so-applied method resulted in 1478 F g^{-1} of ultrahigh specific capacitance when the material was tested in 1 M KOH aqueous electrolyte (H. Y. Wu and Wang 2012).

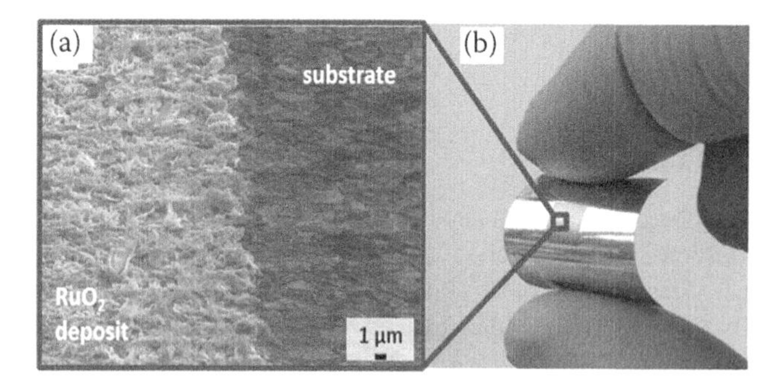

FIGURE 4.6 (a) SEM image of RuO_2/substrate interface of Kapton™/Ti/Au/RuO_2 electrode material and (b) demonstration of laser coated RuO_2 on flexible substrate.

Source: Reprinted with permission from (K. Brousse et al. 2018), Copyright © 2018 Elsevier.

4.2.3 Manganese Oxide

In recent years, MnO_2 has garnered huge research interest attributed to its distinct physical and chemical properties together with a broad range of applications in catalysis, ion exchange, energy storage, biosensor, and molecular adsorption. Special attention has been dedicated to MnO_2 for application as an electrode material for supercapacitors due to its low cost, excellent capacitive performance in aqueous electrolytes, and environmental friendliness (Uke et al. 2017; Iro, Subramani, and Dash 2016).

4.3 POLYMERS

4.3.1 Pure Conducting Polymers

To date, conducting poylmers (CPs) have projected themselves as very promising materials of pseudocapacitive origin attributed to their distinct characteristics. Polyaniline (PANI), polypyrrole (PPy), and olythiophene (PTh) are some of the popular CPs. Supercapacitor electrodes designed with these materials exhibit several advantages, including flexibility, good conductivity, facile synthesis, and cost-effectiveness (Naskar et al. 2021). A substantial quantity of research has been devoted to the electrochemical property improvement of these CP-based electrodes. In this section, research progress of pure PANI, PTh, and PPy based supercapacitor electrodes have been reviewed.

4.3.2 Polyaniline (PANI)

PANI can be synthesized by polymerization of the aniline monomer using different approaches. This material exhibits a lot of advantages, including facile synthesis method, facile doping de-doping related to acid/base chemistry and environmental stability (A. P. Singh et al. 2015), and has projected itself as one of the most promising materials suitable for application as a supercapacitor electrode material. PANI nanostructures' morphology reflects a significant influence on their electrochemical performance. Attributed to this development of a high-efficiency and convenient synthesis approach to synthesize PANI exhibiting necessary nanostructure is very important. In fact, PANI is also capable of synthesizing facially by chemical or electrochemical polymerization. During the chemical oxidative polymerizationin aqueous solution, PANI is mostly positioned as nanofibers (Syarif, Ivandini Tribidasari, and Wibowo 2012). There are several kinds of polymerization approaches to synthesize PANI nanofibers (Abdolahi et al. 2012). Interfacial polymerization is relatively cost effective and facile, making it one of the most common approaches. Sivakkumar et al. (Sivakkumar et al. 2007) synthesized PANI nanofibers *via.* interfacial polymerization approach and evaluated their electrochemical properties through designing a redox supercapacitor tested in aqueous electrolyte in two-electrode configuration. The so-fabricated supercapacitor showed a large initial specific capacitance (554 F g^{-1} at 1.0 A g^{-1}). Nevertheless, the cyclic life corresponding to the material was disappointing, as shown in Figure 4.7.

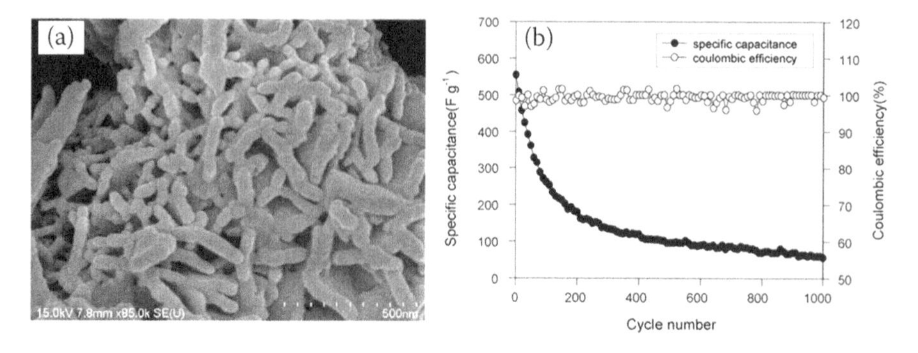

FIGURE 4.7 (a) SEM image of PANI nanofibers and (b) Difference of specific capacitance and coulombic efficiency with respect to number of cycles.

Source: Reprinted with permission from (Sivakkumar et al. 2007), Copyright © 2007, Elsevier.

Li et al. (Hanlu Li et al. 2009) determined the experimental and theoretical specific capacitance of PANI when tested in 1 M H_2SO_4. The greatest theoretical specific capacitance of PANI was obtained to be 2000 F g^{-1} which seemed impressive. Nevertheless, the experimental value of specific capacitance of this material by various approaches was much lesser compared to the theoretical value. This reduced value of the specific capacitance is attributed to contribution to specific capacitance only by a small amount of PANI. The amount of effective PANI contributing to the specific capacitance depends on the material conductivity and also the rate of diffusion of the counter-anions. In a nutshell, pristine PANI implemented as an electrode for supercapacitor application has been investigated significantly, while its electrochemical property, specifically the cycling stability, cannot meet the requisite for the practical applications yet. Deprived cycling stability associated with the supercapacitor results in the reduction of the specific capacitance rapidly, resulting in a few initial cycles. As a result, it has been tried to develop various PANI-based composite with carbon-based materials, metal-based compounds to improve the electrochemical properties of PANI and its cyclic stability (J. Banerjee et al. 2019).

4.3.3 Polypyrrole

Polypyrrole (PPy) is considered among the important CPs due to several advantages, including facile synthesis, comparatively large charge-storage property, and improved cycling stability. Yang et al. (Qinghao Yang, Hou, and Huang 2015) fabricated self-standing films of PPy films using interfacial polymerization in the presence or absence of surfactant. The films synthesized by implementing the surfactant created more pores of smaller size or vesicles and exhibited superior electrochemical performance. For instance, the optimized specific capacitance attained 261 F g^{-1} at 25 mV s^{-1} reserving 75% of its foremost specific capacitance following 1000 cycles, as shown in Figure 4.8. Li and Yang (M. Li and Yang 2015) synthesized an underlying PPy flexible film using an approach of chemical oxidation using methylorange-$FeCl_3$ as the self-sacrificing reactive template. At the $FeCl_3$ to monomer molar ratio of 0.5, the received film consist of nanotubes 5–6 μm

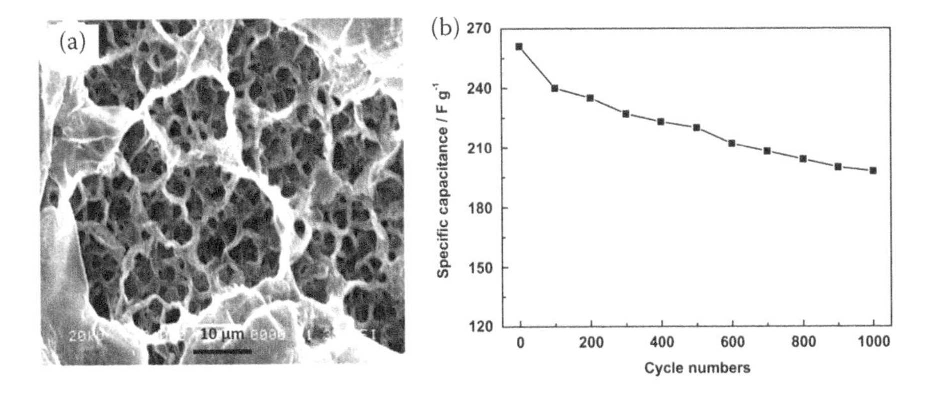

FIGURE 4.8 (a) SEM image of self-standing PPy film synthesized with Tween80 surfactant at 25°C and (b) variation of specific capacitance with cycle life 1 M $NaNO_3$ at 25 $mV.s^{-1}$.

Source: Reprinted with permission from (Qinghao Yang, Hou, and Huang 2015), Copyright © 2014, John Wiley and Sons.

along with a diameter of about 50–60 nm. The PPy film showed excellent electrochemical performance corresponding to the electrochemical charge-storage capacity of 576 F g^{-1} at 0.2 A g^{-1} in combination with an initial capacitance retention of 82% following 1000 cycles at current density of 3 A g^{-1} in 1 M KCl.

Xu et al. (Cao et al. 2018) applied methyl orange (MO) and $FeCl_3$ as a template to synthesize cotton fabrics with good electrical conductivity coated with nanorods of PPy, implementing an *in-situ* polymerization approach. The so-manufactured fabrics were not; when tested as electrodes for supercapacitors, they exhibited a specific capacitance of 325 F g^{-1} and energy density of 24.7 Wh kg^{-1} at 0.6 mA cm^{-2}. The cyclic stability of the device was also tested, and it was found that 63% of the initial capacitance value was retained. Rajesh et al. (Rajesh et al. 2016) synthesized a film comprising PPy using dopant phytic acid through the electro-polymerization method. The maximum specific capacitance obtained for the film was 343 F g^{-1} at 5 mV s^{-1}. Furthermore, the specific capacitance retention of the PPy-based electrode was 91% at 10 A g^{-1} following 4000 cycles. Hence, the synthesis method, dopant, substrate, template, and so on are the significant features affecting the electrochemical character of PPy-based electrodes. Tuning these factors, the electrochemical properties of the PPy can be improved. Therefore, studies on composites of polypyrrole and carbon, polypyrrole and metal based compounds are continued with an aim to improve the electrochemical performance and the cyclic stability. The supercapacitor comprising polythiophene-based electrode exhibit a specific capacitance of 260 F g^{-1} at 2.5 mA cm^{-2}.

4.3.4 Polythiophene

Polythiphene (PTh) and its derivatives have projected themselves as promising materials for application as electrode materials in supercapacitors. PTh have garnered considerable researcher interest due to large environmental stability, high electrical conductivity, and long wavelength absorption (Q. Chen et al. 2019). Significant research has been devoted to investigate the electrochemical characteristics of supercapacitor electrode comprising pristine PTh (Senthilkumar, Thenamirtham, and Kalai Selvan 2011). Several synthesis factors have been tuned to augment the performance of the PTh. Laforgue et al. (Laforgue et al. 1999) synthesized PTh through a chemical approach, and the active material showed a large specific capacitance of 40 mAh g^{-1} and exceptional cycle stability where its capacity was almost the same over 500 cycles. The supercapacitor fabricated with PTh-based electrode showed a specific capacitance of 260 F g^{-1} at 2.5 mA cm^{-2}. Ambade et al. (Ambade et al. 2016) fabricated a wire-shaped and flexible all-solid-state symmetric supercapacitor. In this work, using the electrochemical method, PTh was deposited on titania (TiO_2) wire. The resulting supercapacitor was characterized with a specific capacitance reaching a value of 1357.31 mF g^{-1} along with good cyclic stability corresponding to the maintenance of 97% of the initial capacitance at the end of 3000 cycles. Gnanakan et al. (Gnanakan, Rajasekhar, and Subramania 2009) reported the synthesis of the nanoparticles of pure PTh and the composite of PTh-tartaric acid nano-particles. In the composite, the tartaric acid was implemented as a doping agent possessing cationic surfactant-assisted polymerization approach. The specific capacitance corresponding to the two different types of PTh-based nanoparticles viz composite of PTh-tartaric acid and pristine PTh nano-particles were reported to be 156 and 134 F g^{-1}, respectively. Unsubstituted films of PTh were designed by Nejati, implementing an oxidative CVD approach, and the PTh was then coated on different substrates. Results of the electrochemical experiment indicated that in comparison with the pristine activated carbon electrode, the PTh-coated activated carbon showed better electrochemical properties and improved specific capacitance by 50%. In addition, for the electrodes the cyclic stability corresponded to the 90% of initial capacitance retention following 5000 cycles. Amorphous PTh thin film was synthesized and reported by Patil et al. (Patil, Jagadale, and Lokhande 2012) through undertaking consecutive ionic layer reaction and adsorption approach at ambient temperature. In this method, the oxidation agent used is $FeCl_3$. The thin-film-based electrode when tested as supercapacitor could attain a value of 252 F g^{-1} when measured in 0.1 M $LiClO_4$ solution. Chemical bath deposition method was also undertaken to prepare PTh film and the specific capacitance to obtain for the film 300 F g^{-1} at 5 mV s^{-1} in the solution of 0.1 M $LiClO_4$/PC (Patil, Patil, and Lokhande 2014). In a nutshell, various factors influence the electrochemical properties of the PTh active material, including substrate, synthetic method, morphology of PTh, and so on. Though the performance of PTh-based supercapacitors has been much improved, they are still far from the performance requisite for the practical application. Their electrochemical performance is much inferior in front of the performance of PANI and PPy (Alabadi et al. 2016). Although pristine CPs possess several unique characteristics, these are not apt to be implemented as supercapacitor electrode material attributed to low electronic conductivity and low energy density. For improving electrochemical performance of the CP-based materials, designing of binary and ternary composites of CP has been tried. Carbon materials and metal-based compounds have been used as the fillers to improve the performance of CPs.

4.4 HYBRID MATERIALS

It is well known that transition metal oxides (TMOs) exhibit poor electronic conductivity, reduced specific capacitance, and low stability. The composite synthesized using the carbon material and the TMOs will lead to improvement of the operation of the associated supercapacitors. Hence, in adding to the metal oxides and mixed-metal oxides, various composites of metal oxides, together with carbon material, have also shown a larger surface area as compared to those of the corresponding pristine material (Poonam et al. 2019).

One of the most important advantages of CNTs is that surface functionalization with materials such as conducting polymers and metal oxides can be performed. Hence, there exists a great deal of inquest devoted to contrasting the properties of the pristine materials with nanocomposite. For example, the hydrothermal method was used to synthesis the composite of MnO_2/CNT, and the electrode implementing the nanocomposite exhibited a larger rate capability and specific capacitance as compared to pristine MnO_2 and CNT electrodes. According to the report, the large specific capacitance exhibited by the synthesized nanocomposite was attributed to large specific surface area of the MnO_2 and the large porous structure (F. Liu et al. 2012).

In another work, MnO_2/carbon naospheres (CNS) was reported to be synthesized with very large stability, corresponding to 96.1% of the initial capacitance retention at 5 A g^{-1}. It was recorded that the augmented specific capacitance and cycling life of MnO_2/CNS were primarily due to the robust coupling in in the middle of the nanospheres of carbon material and *in situ* formed sheet arrays of MnO_2 characterized by large specific surface area (L. Wang et al. 2015). Furthermore, various hybrid materials including Co_3O_4/graphene (Xiang et al. 2013) Co_3O_4/CNTs (Shan and Gao 2007), and Co_3O_4/carbon nanofibers (CNFs) (Abouali et al. 2015) have been reported to be synthesized, exhibiting very large surface area along with an augmented conductivity. Besides, an asymmetric supercapacitor machine was built implementing graphene and MnO_2, the cathode was made of the composite of MnO_2-coated/graphene, and the pristine graphene was implemented as the anode. A capacitance value of 245 F g^{-1} at 1 mA is exhibited by pristine graphene electrodes, while a capacitance of 328 F $g^{-1}$is shown by the electrode with deposited MnO_2 at the same current density (Q. Cheng et al. 2011b).

According to a report, in the designing fabrication of a hybrid electrode for ASSCs, hybrid NiO/GFnanocomposite was implemented and displayed 1225 F g^{-1} as specific capacitance at 2 A g^{-1}. An asymmetric supercapacitor was designed, implementing nikel oxide/GF as the cathode and hierarchical porous nitrogen-doped CNTs (HPNCNTs) as the anode; the device was tested in KOH as the electrolyte. This system showed superior electrochemical properties as a result of the established synergy among the two electrodes. The designed device exhibited a large energy density of 32 Wh kg^{-1} at a power density of 700 W kg^{-1} with a 94% of the initial capacitance was retained after 2000 cycles (Huanwen Wang et al. 2014).

Furthermore, $CoNi_2S_4$/graphene (CNS/GR) nanocomposite was reported to exhibit a large specific capacitance of 2009 F g^{-1} at 1 A g^{-1} and the specific capacitance was retained to a value of 755.4 F g^{-1} at 4 A g^{-1} following

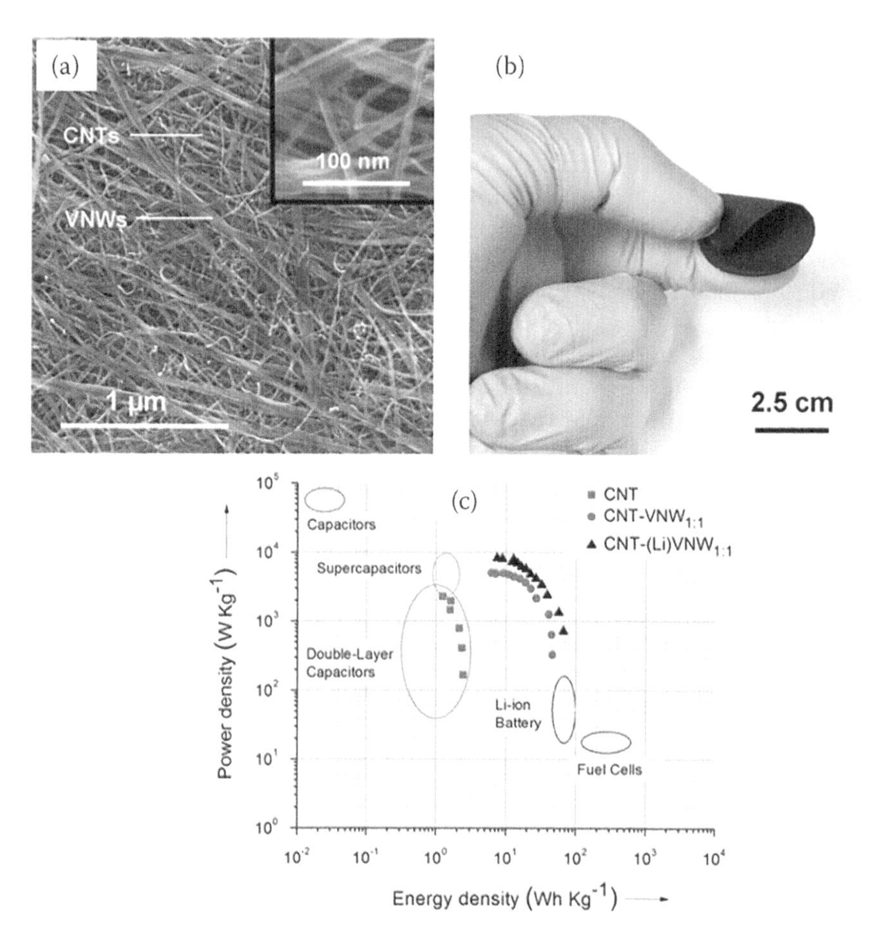

FIGURE 4.9 (a) SEM image of V_2O_5-CNT nanocomposite with marked V_2O_5 nanowires (VNWs) and CNTs (inset image shows nanocomposite at higher magnification), (b) Digital image showing free-standing flexible CNT-VNW paper, and (c) Ragone plot of the CNT-VNW and CNT- (Li)VNW nanocomposite in comparison with other type of energy storage devices.

Source: Reprinted with permission from (Perera et al. 2011), Copyright © 2011, John Wiley and Sons.

2000 charge/discharge cycles (W. Du et al. 2014). In another study, Perera et al. (Perera et al. 2011) designed asymmetric supercapacitor in the coin cell configuration implementing V_2O_5-CNT as the anode and carbon fiber as the cathode (shown in Figure 4.9 a, b). The device delivered 5.26 kW kg^{-1} of power density of and an energy density of around 46.3 Wh kg^{-1} as shown in Figure 4.9 (c). Apart from this, the designed CNTs/nano carbon sphere (NCS) coated by MnO_2 composites exhibited a specific capacitance of 312.5 F g^{-1} at 1 A g^{-1}. Furthermore, this exhibited a capacitance retention of 92.7% following 4000 cycles. The asymmetric supercapacitor design implementing activated carbon as the negative electrode material and the CNTs/NCS/MnO_2 composite as the positive electrode material showed a large specific capacitance associated with a steady voltage window of 1.8 V. The energy density displayed was 27.3 Wh kg^{-1} and the power density was 4500 W kg^{-1} (Lin Hu et al. 2012).

Another substitute is the carbon-based composites of RuO_2 where the content of the carbon materials and RuO_2 including carbon aerosols, AC, graphene and CNTs will render large charge-storage properties and the composite will also be cost effective as compared to the devices implementing pristine RuO_2 based electrodes (D. Hong and Yim 2018). By the introduction of RuO_2 into MWCNTs, MWCNT/RuO_2 composite was synthesized, and the associated device, when tested in acidic solution, for various loading of the RuO_2 filmsexhibited a maximum specific capacitance of 628 F g^{-1} (J. K. Lee et al. 2006).

As mentioned before, CNTs can act as a scaffold to the hollow-structured particles. As an example, one of the leading issues related to the electrodes based on RuO_2 are developing cracks in the body of RuO_2 causing bad cyclic stability. This degradation of the structure is caused by the developed strain as a result of repetitive charge–discharge cycles. For mitigating this issue, a core–shell-templated method was implemented by Wang et al. (P. Wang et al. 2015) for assembling CNT-scaffolded nanoparticles of hollow-structured RuO_2 (hRuO_2/CNT), where it was possible to refrain from the crack formation. The so-designed electrode exhibited a specific capacitance of 655 F g^{-1} at 5 A g^{-1}. Guan et al. (Guan et al. 2014) designed the needle-like Co_3O_4 grown on graphene used as supercapacitor electrode material. The resultant electrode materials exhibited a specific capacitance of 157 F g^{-1} at 0.1 A g^{-1}. In addition, there are many instances where the single-phase materials or the nanocomposites of $Ni_xCo_{3-x}S_4$ could be designed and exhibited excellent specific capacitance. For instance, according to a recent report, $NiCo_2S_4$ nanotubes were designed by implementing sacrificial templates, which exhibited specific capacitance of 933 F g^{-1} (Wan et al. 2013). Ternary $NiCo_2S_4$ delivers greater redox reactions as a result of having a larger electronic conductivity as compared to the $Ni_xCo_{3-x}S_4$ as a result of the reduction in the charge transfer resistance. As a consequence, a little interior resistance (IR) reduction takes place at higher current density. Hence, a larger power density and a higher rate capability was exhibited by the fabricated device (Zhi et al. 2013a). Besides, by a simple hydrothermal Ni–Co–Mn, nanoneedles were designed and the synthesized material showed a specific capacitance of 1400 F g^{-1}, larger energy density, and power density (30 Wh kg^{-1} and 39 kW kg^{-1}, respectively), as compared to Ni–Co–S electrodes. According to the report, there was no loss of the initial capacitance till 3000 charge discharge cycles (Xiong et al. 2015).

Carbon-based materials implemented as the supercapacitor electrodes can be compared based on several parameters, including pore features, surface area, cyclic stability, and cost. The carbon materials and the composites containing metal oxides project themselves as strong candidates as electrode materials (Zhi et al. 2013b). For example, a ternary nanocomposite has been reported to be synthesized implementing the rGO nanospheres and MnO_2 nanorods, which were grown on the poly(3, 4-ethylenedioxythiophene)-poly(styrenesulfonate) (PEDOT: PSS). For supercapacitor application, the synthesized composite (denoted as MGP) exhibited an augmented specific capacitance with stable charge-storage capacitance of 100% following 5000 cycles. The augmented capacitance was a result of association of both double-layer and pseudocapacitive mechanisms (Hareesh et al. 2017). In addition, W. Peng designed composite $NiCo_2S_4$/rGOof nanospheres through hydrothermal method. Specific capacitance of 1406 F $g^{-1}$was exhibited by the resulting composites in combination with a cyclic stability of 82.36% following 2000 charge/discharge cycles at 1 A g^{-1}. These values were greater than the values obtained for pristine $NiCo_2S_4$ (W. Peng et al. 2020).

In another study, composite of carbon and iron cyclotetraphosphate ($Fe_2P_4O_{12}$ was implemented as a supercapacitor electrode material. The authors stated that the phosphorus doping in carbon with porosity and the existence of functional groups in the carbon derived from phytic acid caused improved redox couple, in turn maximizing the pseudocapacitive performance of the composite. The authors have reported that the charge-storage property and the cyclic stability of the resulting composite is not only influenced by the electrode material but also significantly influenced by the selection of the electrolyte. $Fe_2P_4O_{12}$ composite showed a large stability when measured in 0.5 M H_2SO_4. It exhibited a capacitance of 251 F g^{-1} at 1 A g^{-1} accompanied by a constant capacitance following 9000 cycles (Soni and Kurungot 2019).

Furthermore, W. Peng designed the composite of $NiCo_2S_4$/rGO nanospheres *via* hydrothermal approach. The so-obtained composites exhibited a specific capacitance corresponding to a value of 1406 F g^{-1} in combination with the capacitance retention of 82.36% over 2000 charge/discharge cycles at 1 A g^{-1}. These values were greater as compared to the pristine $NiCo_2S_4$ material (W. Peng et al. 2020).

In another study, iron cyclotetraphosphate ($Fe_2P_4O_{12}$) has been composited with carbon was applied as supercapacitor electrode material. According to the reports, the existence of the functional groups in the phytic acid-derived carbon and the phosphorus doping in the porous carbon structure were mainly credited to the increased electrochemical properties of the redox couple. They have stated that the behavior of charge storage and the stability of the composite are influenced not only by the electrode material choice but also by the selection of the electrolyte. $Fe_2P_4O_{12}$ composite showed a specific capacitance of 251 F g^{-1} at 1 A g^{-1} accompanied by a constant capacitance followed by 9000 cycles (Soni and Kurungot 2019).

4.4.1 Carbon-CPs Composites

Various carbonaceous materials can be combined with conducting polymers to result into a nanocomposite. For example, the nanocomposite synthesized by combining a carbonaceous material and conducting polymer can exhibit energy density greater than pristine carbon and power density greater than the pristine conducting polymer. Implementing CPs as the anode and activated carbon as the cathode can result in higher energy density and power density as compared to EDLCs and an improved cycling stability as compared to pseudocapacitors (S. Banerjee and Kar 2020). For an example, a composite electrode was reported to be designed *via.* colloidal self-assembly approach by implementing graphene-MWCNT-PPY nanofibers (Forouzandeh, Kumaravel, and Pillai 2020). The self-standing PEDOT-PSS/SWCNTs composites exhibited a specific capacitance equivalent to 104 F.g^{-1} at 0.2 A g^{-1}, a power density of 825 W kg^{-1}, and an energy density of 7 Wh kg^{-1}. Additionally, there was 90% capacitance retention over 1000 cycles (Antiohos et al. 2011). According to another report, PEDOT/PSS and MWNT composites exhibited 100 $F.g^{-1}$ of specific capacitance (Frackowiak et al. 2006).

Numerous kinds of electrodes can be designed to implement polyaniline (PANI). Nevertheless, the reduced rate capability and poor cyclic stability restrict their applications. In spite of that, materials based on single-phase PANI have been widely believed as a suitable supercapacitor electrode material. There are several nanocomposites that can be designed by combining carbon-based materials and PANI nanostructures showing an improved electrochemical property as compared to that of the parent materials. The composite material containing PANI and carbon scaffolds carbon acts as stabilizer with availability in various morphologies (Panbo Liu et al. 2019). For instance, PANI composite with rGO aerogel was reported to be synthesized by facile electro-deposition approach where rGO aerogel was deposited with the PANI arrays. In general, the deprived energy density, cyclic stability, and rGO aerogels' flexibility limit its application. The synthesized composite exhibited a deep network of open pores and high capacitance value credited to the capacitance originating from PANI and a large electrical conductivity arising from the 3D aerogel, which is having cross-linked framework (Yu Yang, Xi, et al. 2017). The specific capacitance of the resultant composite was reported to be 432 $F.g^{-1}$ at 1 $A.g^{-1}$. The energy density reached a value of 25 $Wh.kg^{-1}$, and a capacitance retention of 85% was shown at the end of 10,000 charge/discharge cycles (Yu Yang, Xi, et al. 2017). In another study, a flexible supercapacitor was designed by implementing composite of etched-carbon fiber cloth and PANI. The material exhibited a capacitance retention of 88% and a specific capacitance reaching a value of 1035 $F.g^{-1}$ at 1 $A.g^{-1}$) (P. Yu et al. 2013). The one-step *in-situ* polymerization approach was undertaken to synthesize the graphene/PANI nanofiber composite, which exhibited a specific capacitance of 526 $F.g^{-1}$ at 0.2 $A.g^{-1}$ (Mao et al. 2012). The so-synthesized GNSs/PANI composite exhibited a specific capacitance equal to 532.3 $F.g^{-1}$ at the scan rate of 2 $mV.s^{-1}$. The composite exhibited the initial capacitance retention of 99.6% at the scan rate of 50 $mV.s^{-1}$ (Ronghua Wang et al. 2017). In addition, in combination with PANI, a CNT hydrogel was designed; it showed a specific capacitance of value 680 mF cm^{-2} at the current density of 1 $mA.cm^{-2}$ (S. Zeng et al. 2015). In another study, PANI nanorods in combination with graphite nano sheets exhibited a specific capacitance of 1665 $F.g^{-1}$ at 1 $A.g^{-1}$ (Yingzhi Li et al. 2013). Han et al. (Zhe Yang et al. 2010) designed electrodes implementing composite comprising graphene oxide and the conducting polymer PEDOT/PSS. When tested in the 1 M H_2SO_4 electrolyte, it exhibited a specific capacitance of value 108 F g^{-1} along with a specific capacitance retention of 78% following 1200 charge/discharge cycles. In addition, the asymmetric (PPy/AC) and symmetric (PPy/PPy) supercapacitor devices were designed implementing Cladophora algae-derived cellulose binder. The devices exhibited a specific capacitance in the range from 0.45 F to 3.8 F (Keskinen et al. 2015). Three separate composites have also been reported to be synthesized *via. in-situ* polymerization by applying CNTs, graphene nano sheets (GNSs), and PANI. The composite of GNSs/PANI showed a specific capacitance of 1046 $F.g^{-1}$. The specific capacitance corresponding to CNT/PANI and GNSs/CNT/PANI composites were 780 $F.g^{-1}$ and 1035 $F.g^{-1}$. The capacitance degradation was 67% for CNT/PANI, 6% for GNSs/CNT/PANI, and 52% GNSs/PANI over 1000 cycles (J. Yan et al. 2010). According to another report hydrogen exfoliated GNSs and polyphenylene diamine showed 248 $F.g^{-1}$ of specific capacitance at 2 $A.g^{-1}$, and the assembled device delivered an energy density of 8.6 $Wh.kg^{-1}$ and a power density of 0.5 $kW.kg^{-1}$ (Jaidev and Ramaprabhu 2012). In another study, graphene-based PVA composites were reported to exhibit 10 folds augmentation in the Young modulus along with a tensile strength 150% improved associated with graphene loading of 1.8 vol % (Xin Zhao et al. 2010). The so-designed composite comprising PEDOT/PSS and CNTs exhibited a specific capacitance ranging from 85 $F.g^{-1}$ to 150 $F.g^{-1}$ at various contents of PEDOT/PSS and CNTs. The energy density was reported to be 0.92 $Wh.kg^{-1}$ while the power density differed from 100 W kg^{-1} to 3000 W kg^{-1} (Snook, Kao, and Best 2011).

A hybrid composite comprising of 2D MoS_2 nanosheets combined with 1D PANI nanowires were implemented as supercapacitor electrode by Nam et al. (Nam et al. 2016), as shown in Figure 4.10 (a). The MoS_2/PANI hybrid electrode material

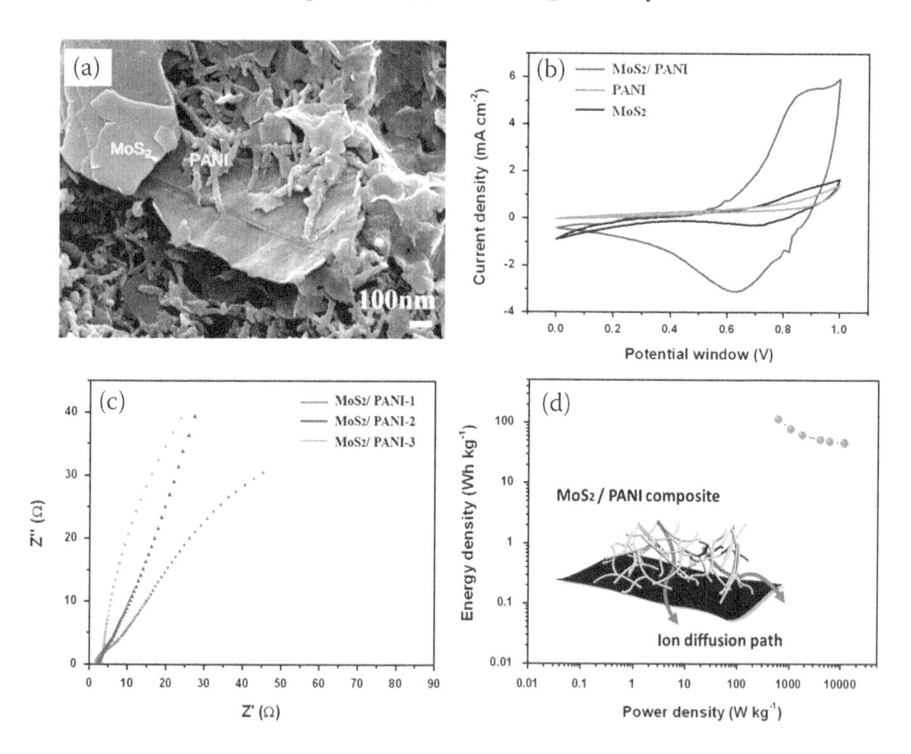

FIGURE 4.10 (a) FESEM image of MoS_2/PANI nanocomposite material, (b) CV curve of pure MoS_2, PANI, and MoS_2/PANI electrode at 20 mV s^{-1} scan rate in aqueous 1 M Na_2SO_4, (c) Nyquist plot from EIS analysis, and (d) Ragone plot of energy density and power density for the MoS_2/PANI electrode.

Source: Reprinted with permission from (Nam et al. 2016), Copyright © 2016, Royal Society of Chemistry.

exhibited a capacitance of about 485 F g^{-1} at 1 mA cm^{-2}, which was higher than pure materials, as can be seen from CV curve in Figure 4.10 (b). The improved specific capacitance reached a value of 812 F g^{-1} for the composite with MoS_2 and PANI in the ratio of 1:2. This improved electrochemical performance was attributed to the established synergy between 1D PANI and the conductive 2D MoS_2 nano sheets exhibiting a high surface area. The Nyquist plot and Ragone plot shown in Figure 4.10 (c) and (d) confirms the high electrochemical performance characteristics ofMoS_2/PANI hybrid electrode material (Nam et al. 2016).

4.4.2 Metal Oxide and Conducting Polymer-Based Composites

The metal oxide (MO) and conducting polymer (CP) based composites (MOs/CPs) are another class of composites with extremely promising electrochemical performance; they can be implemented as electrode material for supercapacitor application resulting from the established synergy between the parent materials (Xie and Wei 2014). The cyclic stability, specific capacitance, and the rate capability can be effectively improved as compared to the electrode material comprising pristine CPs and pristine MOs for electrodes, by creating the composites of CPs with MOs causing improved conductivity of the electrodes. For example, the MoO_3 coated by PPy was reported to be designed *via.* an *in-situ* polymerization approach, which exhibited specific capacitance of 110 F g^{-1} at 100 m A g^{-1}. The energy density obtained was 20 Wh kg^{-1} corresponding to a power density of 75 W kg^{-1}. Asymmetric supercapacitor devices assembled implementing PPy/MoO_3 composite as the positive electrode and AC as the negative electrode when tested in 0.5 M K_2SO_4 aqueous electrolyte exhibited an energy density of 12 Wh kg^{-1} at Power density of 3 kW kg^{-1} (Yu Liu et al. 2013).

The nanocomposite of PANI/MnO_2 was synthesized by the exchange reaction between MnO_2 and n-octadecyltrimethylammonium-intercalated and PANI in N-methyl-2-pyrrolidone solvent (X. Zhang et al. 2007). The Fourier transform infrared spectroscopy results revealed the interactions that developed within the intercalated PANI and the MnO_2 layers. Even after the intercalation into the MnO_2 layers, PANI exhibited considerably good electrical conductivity. The composite was characterized with a specific capacitance of 330 F g^{-1} at 1 A g^{-1}. The improved specific capacitance in the composite is due to developed synergistic effect between MnO_2 and PANI. The initial specific capacitance retention was 94% of the following 1000 charge/discharge cycles. The composite of MOs-CPs could be implemented widely as non-flexible and flexible planar SCs electrode material to optimize various parameters, including the energy density, capacitance, and power density. For instance, Raj et al. (Raj, Ragupathy, and Mohan 2015) fabricated (cobalt oxide-conducting polyindole) Co_3O_4-PIND, and applied that as an electrode material in a non-flexible supercapacitor. The supercapacitor fabricated by implementing a system of free-binder exhibiting a specific capacitance of 1805 F g^{-1} at 2 A g^{-1}. Several composite-based electrode materials, comprising carbon-based material, metal oxides, and conducting polymers, where the best electrochemical property exhibited has been exhibited by the ternary composite as compared to those of parent components (Forouzandeh, Kumaravel, and Pillai 2020).

4.5 FUTURE MATERIALS

Pseudocapacitors are generally known to implement fast and highly reversible redox interactions with the electrolyte at the interphase of the electrode and the electrolyte. The large (as compared to the supercapacitors based on EDL mechanism) specific capacitance of faradaic electrodes garners much attention of the researchers working in the field of energy-storage materials. Several new materials have been discovered and investigated for checking the aptness of a supercapacitor electrode material. For instance, Yafei et al. synthesized two different electrode materials by simple hydrothermal procedure: nanoclusters of zinc–cobalt oxide and sulfide hybrid (ZCOSH) and drew their comparison with binary oxide nanosheets zinc–cobalt (ZCO NSs). Nanoclusters of ZCOSH were designed by fabricating a Ni foam, activated carbon electrodes, and an NKK MPF30AC 100 membrane as a separator. The specific capacitance was determined to be 2176.7 F g^{-1} for ZCOSH nanoclusters and 367.2 F g^{-1} for the ZCO NSs (Najib and Erdem 2019).

5 Macro Supercapacitor

5.1 FLEXIBLE SUPERCAPACITORS (FSCS)

Flexible electronic gadgets are widely applied for daily life requirements, various industrial production uses, and application in aerospace, medicine, military, environmental monitoring, etc. These require perpetual supply of power. Fiber-like flexible supercapacitors (FSCs), also termed as wire-shaped FSCs, are efficient energy providers for the aforementioned applications (Xianfu Wang et al. 2014). Apart from acting as an energy source for micro-electronics, fiber-like FSCs also prove suitable for application in the field of smart clothes (Xianfu Wang et al. 2014). Based on current established weaving knowledge, yarns with apt mechanical strength and flexibility may be fabricated into different clothes with ease (Xianfu Wang et al. 2014).

As shown in Figure 5.1(a), there can be two kinds of structures associated with the fiber-like FSCs. The first type of structure is obtained from fiber-like flexible electrodes (FEs) enfolded around each other, and the second type of structure is associated with a coaxial structure. In between the electrodes generally exist a separator and an electrolyte in an FSC. A separator is a must in case of a liquid electrolyte; however, for solid gel, an electrolyte separator is not necessary. In the second case, the electrolyte itself plays the role of a separator and stops the short circuit (C. Choi et al. 2014). Solid gel electrolytes are readily available and very safe to implement, more so than the liquid electrolytes in FSCs because the liquid electrolyte is challenged by the risk of leakage of the electrolyte. Fiber-like FEs possess significant flexibility and exhibit good electrochemical properties. The characteristics of large flexibility needs FEs with fiber-like properties to show a flexible mechanically and strong fibrous structural scaffold, while the electrochemical properties of the FEs are due to implementation of electrochemically active materials (C. Choi et al. 2014). For example, flexible fiber scaffolds are covered with electrochemical active materials to design FEs, as illustrated in Figure 5.1 (b). There are various types of alternate fibrous structural scaffolds and electrochemically active materials, such as activated carbon, carbon nanotubes, nanosheets of graphene, MnO_2, and polyaniline, etc. In addition, several complex methods have been implemented directly to attain the process of coating, including electrodeposition, chemical reaction "dipping and drying," and so on (P. Yang et al. 2014). However, note that in case of the FEs, as shown in Figure 5.1(b), the volume or/and weight fraction associated with the fibrous structural supports is often much larger compared to electrochemically active materials. Negative effect is imparted as a result of this situation on the specific capacitance and the associated electrochemical peerfotmance of FEs with fiber-like properties (Dong, Xu, Li, Wu, et al. 2016).

DOI: 10.1201/9781003174554-5

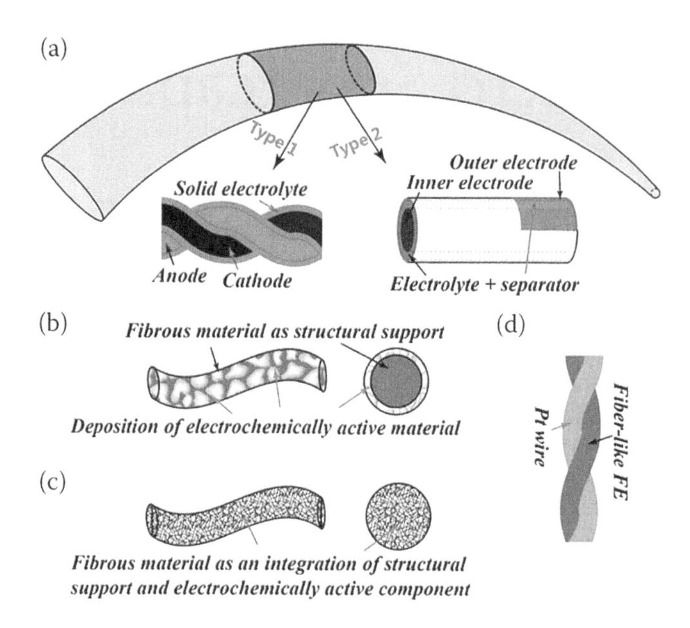

FIGURE 5.1 Schematics of different fiber-like FSCs (a) and FEs (b)–(d) (Dong, Xu, Li, Huang, et al. 2016).

Obviously, for improving the electrochemical characteristics of FEs, it is better to minimize the implementation of materials not exhibiting electrochemical properties. It is significant to develop proper fibrous materials that can function as the structural support and also exhibit electrochemical activity (Figure 5.1 (c)). Furthermore, current collectors form the other important part of a supercapacitor. As shown in Figure 5.1 (d), largely conductive metal wires/meshes (e.g., Pt wire) are mostly implemented in fiber-like FSCs warped with FEs as current collectors (Foroughi et al. 2014). However, with exception, the requirement for the current collector is eliminated in case the electrical conductivity of the electrochemically active material is meaningfully good (C. Choi et al. 2014). To further study the above concept, representative researches employing FSCs along with fiber-like FEs are discussed.

5.1.1 FEs/FSCs with Plastic Fiber Scaffold

Inexpensive, mechanically robust and flexible plastic fibers are possible to be applied as fibrous structural scaffold in FEs with fiber-like properties. In accordance with a report by Fu et al., plastic fiber was covered through a method of dipping used with commercial pen ink (Y. Fu et al. 2012). The solid part existing in the pen's ink, as shown in Figure 5.2 (a), was basically graphite carbon nanoparticles. The nanoparticles from the pen ink resulted in a uniform film on the plastic fiber (Figure 5.2 (b)). Besides having high electrical conductivity, they acted as electrochemically active materials in FEs and also acted as current collector. Beforehand, Au film was grown on the plastic fiber surface. The assembled FSC implementing plastic fiber

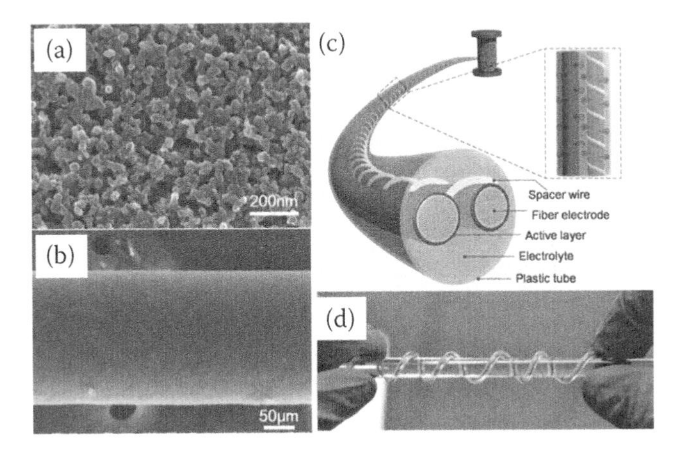

FIGURE 5.2 SEM image of the electrochemically active materials (a) and fiber electrode made of plastic (b); schematic of structure (c) and digital picture (d) of symmetric FSCs with fiber like properties.

Source: Adapted from (Y. Fu et al. 2012) with permission from Wiley-VCH.

electrodes/pen ink (Figure 5.2 (c) and (d)) exhibited a capacitance, energy density, and power density of 19.5 mF/cm^2, 2.7 mW.h.cm^{-2} and 9070 mW.cm^{-2}, respectively. The FSC exhibited good flexibility and cyclic life. It was observed that the absence of any decay of the initial capacitance following 15,000 cycles tested through cyclic voltammetry measurements. Furthermore, with bending the FSC from 0° to 180° and 360°, only a very small reduction in the capacitance value was noted. Furthermore, it is noteworthy that in addition to the use of PVA/H_2SO_4 gel electrolyte, for maintaining a gap between the two electrodes, implementation of a wire (Figure 5.2 (c)) was designed to refrain from the direct contact between two fiber-like electrodes.

5.1.2 Natural Fiber Supported FEs/FSCs

Yarn made up of cotton, linen, and bamboo, etc., exhibits similar diameter and a geometric structure of the yarn bundle: each yarn possesses several individual mono filaments with diameter around 10 nm. Meanwhile, attributed to the flexibility of the above yarns and their robustness, they can be implemented as scaffolds for the fibrous structures in the fiber-like FEs (Jost et al. 2015). Unlike the largely conducting carbon-fiber (CF) bundles, these yarns generally are insulating in character. Therefore, it is requisite to implement conductive fillers such as CNTs, graphene nanosheets, and activated carbon particles for fabricating yarn-based fiber-like FEs.

5.1.3 FEs/FSCs Scaffolded by CNT Yarn

CNT yarns, graphene nanosheet (GN) fibers, and GN/CNT hybrid fibers exhibit multiple advantages, including large electrical conductivity, mechanical strength, lightweightness, and relatively large specific surface area (Meng et al. 2013).

Besides, GN and CNTs exhibit electrochemical performances by themselves. Hence, it is appropriate to keep in place CNT yarns, GN fibers, or their hybrid to design FEs with fiber-like properties and FSCs. By winding CNT films, CNT yarns are normally fabricated through CVD method (C. Choi et al. 2014). Smithyman et al. applied CNT yarns as FEs with fiber-like properties (Smithyman and Liang 2014) and demonstrated that the capacitance came around only 20 $F.g^{-1}$ at 50 $mV.s^{-1}$. An assembly is created from a coaxial fiber-like FSC having CNT yarns as the inner electrode and an outer electrode of CNT film. Some other value of capacitance is reported in other works using CNT yarns, including SWNTs and MWNTs; however, overall, the capacitances obtained for CNT yarns are not good enough (Y. Huang, Hu, et al. 2015). Actually, the so-produced yarns made from CNTs are relatively unsuitable for ion diffusion, specifically in solid gel electrolytes (Foroughi et al. 2014). This is a significant reason detracting from the obtained electrochemical capacitance associated with the CNT yarn electrodes. In addition, partial elimination of the catalysts and lack of a surface-activation procedure also affect negatively the electrochemical performance (Smithyman and Liang 2014). Though CNT yarns are possible to be implemented in FEs with fiber-like properties, but these are unable to provide large electrochemical capacities. Hence, it is still required to implement electrochemically active materials into the CNT yarns for producing composite yarn electrodes with high performance.

5.1.4 Paper-Like FEs and FSCs

Paper-like FSCs, also termed as planar FSCs, are implemented as electronic screens associated with digital cameras, foldable mobile phones, and laptops. These are integrated along with a current collector, paper-like FEs, a flexible packing shell, and a separator (S. Y. Lee et al. 2013). In spite of the newer concept regarding FSCs with paper-like features, paper-like FEs have been in the market for a long time. For instance, for powder-like electrochemically active materials, they are usually covered on a metal foil, including flexible Cu foil, Al foil, or Ni foil, for electrochemical measurements. From a particular perspective, in this moment, metal foil covered with the active-material is taken as a paper-like FE. According to the implementation of substrate, the reported paper-like FEs are classified into two families, freestanding FEs and flexible substrate scaffolded FEs. In case of the free-standing FEs, there is the presence of robust network structures. Penetration of the electrolyte into the inner layers of the electrode depends on the pores defined by network skeletons. Normally speaking, metal oxide/hydroxide particles are very fragile and cannot be cast into flexible self-standing networks; furthermore, as the particles own branch-like or wire-like morphology, which can be synthesized successfully, such materials are rarely reported. CNFs and CNTs display a high ratio of length-to-diameter, adequate flexibility, large mechanical strength, and significantly high electrical conductivity (X. Yan et al. 2011). As a result, flexible, free-standing, and conductive carbon films are found, which can be straight away applied as paper-like FEs, even though their electrochemical performance depends on several factors, inclusive of their specific surface area and electrical conductivity. The electrochemical charge-storage capabilities of the paper-like FEs will additionally enhance the incorporation of some other electrochemically active materials, including electrically conducting polymers and metal oxides/hydroxides. Graphene nanosheet in combination with electrically conducting polymers exhibiting long chain structures is also possible for fabrication into free-standing film FEs (Dikin et al. 2007). It is ostensible that self-supporting FEs refrain from the requirement of extra substrates and/or current collectors when assembled into FSCs with paper-like features, largely reducing the mass of electrodes and hence supercapacitors. However, for some different electrode materials, such as spherical AC particles, and metal oxides/hydroxides assembling them into freestanding FEs becomes very difficult. Attributed to this, paper-like FEs are mostly fabricated with the help of electrochemically active material deposition on suitable flexible substrates (S. Chen et al. 2010).

5.1.5 3D Porous FEs and Corresponding FSCs

Several fiber-like and paper-like FSCs exhibit excellent energy densities, power densities, and gravimetric capacitances. They are suitable for application in microelectronic devices. However, they are very tough to familiarize for large-scale equipment, which will require large energy over a short time (Zhai et al. 2015), and the power of a paper-like FSC or a single fiber-like FSC is very small. Interlacing different fiber-like FSCs in one big textile supercapacitor can aid reaching high-energy density (Meng et al. 2013), but this still is challenged by various drawbacks that must be mitigated. The glitches include fabrication of a robust enough, long enough, and cost effective fiber-like FSCs for implementation for the braiding method. Similarly, the outstanding properties of paper-like FEs is always influenced by the electrode materials' narrowness, which results in less areal energy density. However, improving the thickness of the electrode material will severely damage the gravimetric capacitances of electrodes/supercapacitors (Shah, Zhang, and Talapatra 2009). In addition, a few excellent-performance paper-like and fiber-like FSCs are reported, which were synthesized by implementing costly raw materials and/or complex fabrication methods, highly hindering their applications in real life. For fabricating supercapacitors with large energy output, porous and broad electrodes (3D electrodes) developed (Jost, Dion, and Gogotsi 2014). Each porous 3D electrode consists of a porous framework and an electrochemically active filler. For 3D porous electrodes with flexibility, comprising of aerogel electrodes, flexible fabric electrodes, and electrodes like sponge, and so on, the host materials are broad (~100 micrometers), porous, and flexible. These properties are helpful in these ways: (1) creating a good flexibility accompanying the 3D porous electrodes; (2) creating space for the addition of a high loading quantity of fillers comprising electrochemically active material in main materials, rendering it possible to synthesize high-energy electrodes; and (3) promoting fast transportation of ions in the interior of the electrodes and hence preferring a large power density. It is evident that the body material will mainly function as a flexible frame scaffold, and the electrochemically active material filler is responsible for delivering the electrochemical capacity. However, in a few cases, the two parts, particularly the former ones, can also act as other materials in 3D porous FEs.

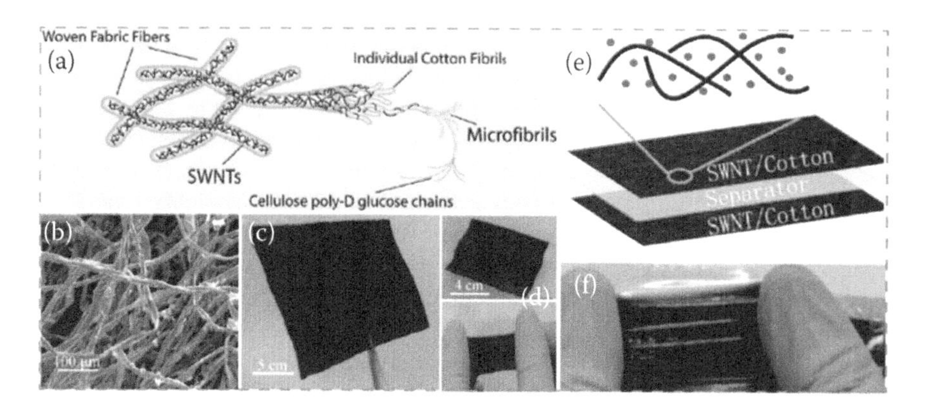

FIGURE 5.3 (a) SEM (b) and digital pictures images (c) and (d) of stretchable flexible electrodes; fabrication (e) and photo (f) of stretchable fabric supercapacitors.

Source: Reproduced from ref. (Liangbing Hu et al. 2010) with permission from the American Chemical Society.

5.1.6 Textile FEs and FSCs

Textiles are illustrative body of resources applied in 3D porous FEs. Polyester microfiber twill, carbon foam fabrics, and cotton cloth, etc., have been applied as such textiles (Jost, Dion, and Gogotsi 2014). These textiles possess a similar preparation method and structure (textile technology) like the mostly used materials for clothing. Hence, textile FEs/FSCs can be considered for construction of smart garments (Jost, Dion, and Gogotsi 2014).

Hu et al. reported the designing of stretchable textile-based electrodes by the integration of SWNTs into cotton cloth (Figure 5.3 (a)–(f)) (Liangbing Hu et al. 2010). The SWNTs powerfully got stuck to the cotton fibers in textiles, hence limiting their reduction in the stretching and folding of the textiles (Figure 5.3 (d)). The body material of textiles fabricated out of the cotton cloth exhibited electrically insulating properties and hence refrained from exhibiting any electrochemical activities. But due to its thickness and porous structure, it was able to house a large amount of electrochemically active SWNTs loading. This resulted in a improved electrical conductivity for the textile electrodes and an areal capacitance (0.48 F cm^{-2}) of the supercapacitors based on the electrodes (Figure 5.3 (e)). Apart from the electrochemical performances, the textile supercapacitors remained almost the same following stretching, as shown in Figure 5.3 (f).

5.2 OPTICALLY ACTIVE SUPERCAPACITOR

An optically active supercapacitor or photo-supercapacitor is a combination of an energy-storage device consisting of a dye-sensitized solar cell (DSSC), acting as the primary electron contributor attributed to the excitation of the electrons related to the dye and the semiconductor conduction band, and a supercapacitor as the electron storage sink. The DSSC is associated with a significant thickness of metal

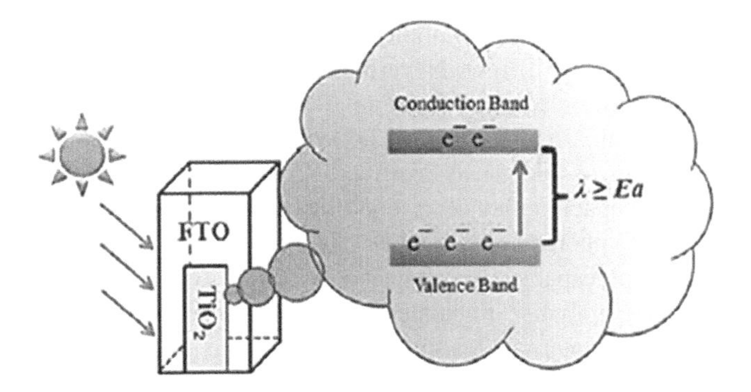

FIGURE 5.4 Transportation of electrons from DSSC compartments to the supercapacitor reservoir.

oxide, which is photoactive on a photoanode substrate; it is a transparent, redox electrolyte, and a counter electrode (Y. Qin and Peng 2012). The commonly implemented parts in a DSSC are titanium dioxide (TiO_2), ITO as a transparent substrate, and Pt and as a counter electrode (Pratiwi et al. 2019). On the other hand, a sandwich-like SC consist of stacked current collectors having electrolyte also playing the role of a separator. Both the devices are composed of the same counter electrode to complete the circuit. A photo-supercapacitor is created to use the renewable and easily available solar energy resource to create a hike in the light-to-electrical energy conversion. A DSSC plays the role of an contributor of electron in light's presence when photons encounter the dye molecules. As the dye is illuminated through the radiation of an equal or greater wavelength than the activation energy of the electrons, a transfer of the electrons belonging to the dye from the valence band highest occupied molecular orbital to the lowest unoccupied molecular orbital belonging to the metal oxide surfaces; it is photoactive, inducing an electron pathway to the supercapacitor reservoir from the DSSC compartments, as shown in Figure 5.4.

The electrons created through the excitation of photons are incorporated into the conduction band belonging to the ITO (~4.7 eV). (Mane et al. 2009). The buildup of holes results in an atmosphere accompanied with significant electron affinity at the sensitizer/electrolyte interface. The I^{3-} redox ions diffusing across the DSSC electrolyte-dye vicinity maintain the photogenerated electron circulation (Uono et al. 2006). Electrons are transported for dye regeneration from the redox electrolyte in the reduced form to compensate the affinity sites (Snaith et al. 2007). At the same time, there occurs transfer of electrons from the counter electrode to provide charge for the vacant sites within the electrolyte (Snaith et al. 2007).

In the case of a photo-supercapacitor, the irradiated electrons are transported from the DSSC over an external circuit and accumulate in the supercapacitor. The discharging and charging of a photo-supercapacitor occurs according to the working principles of an individual supercapacitor, with the only difference that there occurs the utilization of solar energy to initiate the photoelectron generation in place of electrical energy provided by any power source. Photogenerated electrons accumulate in the common counter electrode, which can produce electricity in the absence of light. Nevertheless, the rejoining of redox species or dye molecules and the electrons at the interface of sensitizer and electrolyte is inevitable (Park, Frank, and Korea 2003), leading to a cutoff in the conversion efficiency from light to- electrical energy.

An photo-supercapacitor device is coordinated with a three-electrode device, containing a DSSC photoanode, counter electrode that is shared and a current collector of the supercapacitor in contact with electrolytes. In 2010, a work on PEDOT conducting polymer implemented as electrode in photo-supercapacitor exhibited a smaller internal resistance estimated to be 160 U (Murakami, Kawashima, and Miyasaka 2005). This advantage was attributed to PEDOT, which produced a larger surface area and larger conductivity as compared to the further conducting polymers. The specific capacitance of the photo-supercapacitor was obtained to be 0.52 $F.cm^{-2}$ and the DSSC based on plastic exhibited an efficiency reaching a value of 4.37% (H. W. Chen et al. 2010). In the same year, another photo-supercapacitor based on conducting polymer film was made up from poly (3,3-diethyl-3,4- dihydro-2H-thieno-[3,4-b][1,4]dioxepine) (PProDOT-Et2) for storage of energy and an N_3 dye-TiO_2 DSSC for energy-conversion devices, showing an efficiency of energy storage reaching a value of 0.6% (Hsu et al. 2010). In 2013, a single crystal ZnO (hydrogenated) doped MnO_2 nanoscale-based flexible supercapacitor showing a specific capacitance of 1261 F g^{-1} related with inductively coupled plasma atomic emission spectroscopy as the MnO_2 loading reached a value of 0.11 mg cm^{-2}. In 2014, a photon-to-electrical conversion of 1.64%, and the greatest value of storage efficiency could be attained by a photo-supercapacitor associated with nanotube arrays of bipolar anodic titanium oxide with selective hydrogen plasma treatment on supercapacitor sub-device. This device exhibited a highest energy-storage efficiency of 51.06% and outstanding cyclic stability associated with discharge areal capacitance (1.289 mF cm^{-2}) of 96.5% following 100 cycles at 0.1 mA cm^{-2} (J. Xu et al. 2014). An integrated DSSC-comprising of asymmetric $Ni(Co)O_x$/AC supercapacitor (photo-supercapacitor) reported to exhibit a 46 F g^{-1} of specific capacitance for a single supercapacitor device. For the individual electrodes, in cyclic voltammogram redox, peaks were seen originating from $Ni(Co)O_x$ electrode, representative of the energy storage in the electrolyte/electrode interface ascribed to the redox transition. However, a reduction of 14 F g^{-1} capacitance of the asymmetric supercapacitor was obtained in combination with the DSSC, resulting in a capacitance of 32 F g^{-1} and an efficiency of 0.6%. (Bagheri et al. 2014).

6 Planar Micro-Supercapacitor

6.1 DIFFERENCES IN MACRO-SUPERCAPACITOR AND PLANAR MICRO-SUPERCAPACITORS

The structures of macro-supercapacitor usually comprise planar interdigitated and sandwich-shaped structures, as shown in Figure 6.1(a). This type of difference is also applicable for description of the micro-supercapacitors. In a sandwich-type supercapacitor, the electrolyte lies in between (sandwiched) the two electrodes. At present, the widely developed configuration in case of the macro-supercapacitors is this sandwiched design attributed to its simple design, cost-effectiveness, and possibility to scale up. However, in supercapacitors configured in a sandwich-like fashion, the current passes at 90° to the plane of the device, including electrodes and the electrolyte, as shown at the top of Figure 6.1(b) (J. J. Yoo et al. 2011). The huge resistance coming from the interfaces between the components of the supercapacitor and the large paths for the ion diffusion generally reduced the specific capacitance and power density. Moreover, unwanted dislocation occurring in between layers of the sandwiched supercapacitor is very common, especially when these supercapacitors are bent, making them unsuitable for flexible electronics. On the other hand, planar supercapacitors exhibit various advantages as compared to the sandwiched macro-supercapacitor attributed to their exceptional interdigitated configurations of the electrode (J. J. Yoo et al. 2011; El-Kady and Kaner 2013). To explain, the little interspaces formed in the gap between the interdigitated electrodes lead to reduction in the possibility of electrical short-circuiting and subsequent reduction in the resistance corresponding to transport associated with the electrolytic ions. This improves the electrochemical performance at the electrode–electrolyte interfaces. On the other hand, supercapacitor with the planar configuration are known to reduce the displacement occurring between various layers as is generally observed in the case of the supercapacitors with sandwiched architecture. Planar supercapacitors hence have the advantage of improved mechanical flexibility, significant for application in the field of flexible and portable electronics. Furthermore, there can be substantial improvement in the electrochemical performance at the electrode-electrolyte interspaces as a result of it being interdigitated. This is particularly important for supercapacitors having electrodes made up of 2D materials possessing a thin-film structure along with a planar morphology. This is attributed to the fact that for interdigitated design, it is possible for the electrolytic ions to get transported through the flat electrode surfaces associated with the electrodes based on 2D material bearing a smaller diffusion distance, as has been illustrated in the bottom part of Figure 6.1(b).

DOI: 10.1201/9781003174554-6

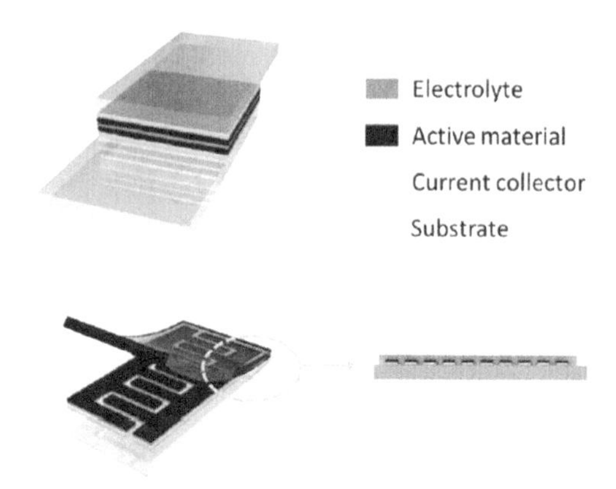

FIGURE 6.1 (a) Sandwich-type (top) and planar SCs (bottom).

Source: Reprinted from ref. (J. J. Yoo et al. 2011) with permission from American Chemical Society, copyright 2011.

It is noteworthy that even though interdigital fabrication is widely implemented in micro-supercapacitors, other different designs, including spiral-shaped electrodes, can result in larger energy/power delivery (X. Tian et al. 2015). It is expected that more advanced electrode designs will prevail in the future given the multipurpose patterning ability of printing. Lastly, for printed supercapacitors, the context of a sandwiched configuration is ineffective as it is not very easy to grow a layer of electrode on a liquid or gel electrolyte. Unlike this method, planar-micro supercapacitors project themselves as a promising technology that can be integrated in a facile way onto the microelectronic devices contained on-chip attributed to their in-plane design and their small sizes. Thus, this is very interesting and also of a great challenge to design interdigitated electrode structures for application in high-performance micro supercapacitors. Till now, various processes have been implemented, such as vacuum vapor deposition, photolithography, and laser scribing (X. Tian et al. 2015; R. Ye, James, and Tour 2018). Nevertheless, the involved processes are not very straightforward, are expensive, and can't be scaled up. In contrast, printing methods come with large patterning ability and compatibility with roll-to-roll and solution-based methods, thus paving effective and feasible manufacturing methods for the cost-effective and scaled-up production of micro supercapacitors possessing interdigitated electrode design. For instance, a printable and scalable approach has been developed to fabricate interdigitated electrodes comprising hierarchical nano-coral structure for application in flexible microsupercapacitors (Y. Lin, Gao, and Fan 2017). Interestingly, as a result of the implementation of the inkjet-printing approach, outstanding adaptability in precisely and artistically fabricating the patterns could be achieved. It was discovered that the specific capacitance (areal) associated with the micro-supercapacitors could be consequently improved along with the decrease in the interspace existing in-between the two neighboring electrodes. All solid-state flexible micro supercapacitors that were inkjet-printed associated with nano-coral structures could attain large areal capacitances of 52.9 mF cm^{-2}. According to other reports, high-area macro-supercapacitors were designed acting as units of energy-storage adapted into a wearable self-power-driven system. This exhibited the high scalability related to the particular printing approach. Printing approaches including screen printing and 3D printing result in easy fabrication of broader electrodes; hence, it is possible to build 3D supercapacitors having higher energy density and areal capacitance as compared to those of 2D thin-film based supercapacitors having the similar footprint area. The 3D structure is responsible for exposing the external shells of the electrodes into the electrolyte in all the three dimensions, easing/augmenting the charge-storage approach, causing higher active surface area as well as enhanced utilization of the active mass. Furthermore, the current collector associated with a porous structure expanding in three dimensions might also cause lesser resistance and hence augmented electrochemical performances. The thick electrodes (> 5 mm) associated with 3D supercapacitors may result in the lowering the power density; nevertheless, the significantly enhanced energy density is normally of greater importance to 3D supercapacitors for application in energy-storage devices. Furthermore, broad but porous electrodes show significant promise of realizing simultaneous increase in the power as well as energy densities. Other than augmenting the energy density and the areal capacitance, the 3D structure proves to be promising in endowing supercapacitors with different functionalities. According to an example, Lv et al. attempted the designing of a stretchable 3D supercapacitor proved to be inspiring using a honeycomb lantern (Y. Lin, Gao, and Fan 2017). The supercapacitor is associated with honeycomb composite electrodes, which are expandable and the ion transport paths are independent of the device-thickness. The 3D supercapacitor exhibits almost 60 folds larger capacitance corresponding to a value of 7.34 F cm^{-2} as compared to its 2D counterpart (120 mF cm^{-2}). Furthermore, the structure expanding in 3D can significantly enhance the stretchability of the supercapacitor enabling customizable shapes, facile integrability of the supercapacitor with wearable/stretchable electronics. Supercapacitors are also generally classified as asymmetric and symmetrical supercapacitors based on the similarity or difference in the composition of the two electrodes. In the case of the symmetrical supercapacitors, both the electrodes consist of similar kind of material with same mass loading either material of electric double-layer origin or pseudocapacitance origin. Asymmetric supercapacitors are the ones where the two electrodes are made up of different materials. Hence, asymmetric supercapacitors exhibit the potential of being associated with large power and energy densities synchronously. These also exhibit good cycling stability together with broad operating potential windows. The significant diverse advantage of approaches involving printing approaches is that different materials can simply be deposited having various mass loadings/thicknesses on substrate. This promises for the fabrication of asymmetric supercapacitors having the face of optimization of the two electrodes. According to a report, the pioneering work implements inkjet printing to design asymmetric supercapacitors grown on a flexible polyethylene terephthalate substrate (Dinh et al. 2014). According to the report, flexible all-solid-state asymmetric supercapacitor implementing graphene nanosheets and nanocrystal whiskers of lamellar $K_2Co_3(P_2O_7)_2.2H_2O$ were effectively fabricated through the method of inkjet printing. The micro-devices assembled showed a volumetric capacitance of 6.0 F cm^{-3}, high rate constancy, and high cyclic stability (5000 cycles) associated with the highest energy density corresponding to the value of 0.96 mW h cm^{-3}. This work demonstrated impactful insights associated with the development of micro-supercapacitors for applications in flexible electronics (Y. Z. Zhang et al. 2019).

6.2 MECHANISM OF ELECTROCHEMICAL INTERACTIONS

Electrochemical devices have garnered significant attention in the previous few decades attributed to the fast rate of charge–discharge and expanded cyclic stability (Augustyn, Simon, and Dunn 2014; Hall et al. 2010). In comparison to other energy-storage systems, including batteries, electrochemical capacitors have larger power densities, and these can be charged and discharged within a few seconds (Kelly-Holmes 2016). From the time when General Electric announced the first patent in the context of electrochemical capacitors in 1957 (Becker 1957b), these have found application in several fields, such as power supply and capture, applications in the context of power quality, and backup power (Hall et al. 2010).

6.2.1 Fundamentals of Double Layer Capacitance and Pseudocapacitance

This section includes a discussion about the mechanism of charge storage in various supercapacitors. Recalling the previous discussion, electrochemical capacitors also termed as supercapacitors can be divided into two categories based on the process of energy storage: pseudocapacitors and electric double layer capacitors (Simon, Gogotsi, and Dunn 2014). In EDLCs, there is the formation of two oppositely charged layers at the interface of the electrode and the electrolyte interface due to the charge separation, and the charge is stored in such double layers (Hall et al. 2010). For the pseudocapacitors, electro sorption/ redox reversible reaction occurs at the electrode surface, and the energy is stored. This happens usually in the case of the conducting polymer or transition metal oxide (Augustyn, Simon, and Dunn 2014; B. Wang et al. 2020). Generally, these mechanism are not completely separable and both exist in a supercapacitor.

6.2.2 Charge-Storage Mechanism in EDLCs

6.2.2.1 Basic Difference between the Electric Double-Layer Capacitor, Pseudocapacitor, and Battery-Based on Charge-Storage Mechanisms

The energy storage in EDLCs is by the means of physical surface charge adsorption in absence of any faradaic reactions (Stoller and Ruoff 2010). During discharge/ charge cycles, the charge accumulation in the double layer of Helmholtz (EDL) leads to production of a displacement current. These types of materials can deliver high power density as they can respond quickly to the variation in the potential as well as the physical reaction in nature. EDL capacitors (EDLCs) are able to deliver energy quickly, creating considerable power (Simon, Gogotsi, and Dunn 2014). Nevertheless, as a result of the incarceration of the electrode surface, the total energy stored becomes limited, and way lesser as compared to the pseudocapacitors and batteries. The EDL capacitance is described as follows (Zhang and Zhao 2009; Liu et al. 2018):

$$C_{dl} = \frac{Q}{V} = \frac{\epsilon_r \epsilon_o A}{d} \tag{6.1}$$

In the above equation, C_{dl} is the electric double-layer capacitance contributed by a single electrode, Q denotes the overall charge transported at the operating potential of V, ε_0 is the dielectric constant of vacuum, ε_r is the dielectric constant of the electrolyte, d is the distance between the two oppositely charged layers, and A is the electrode surface area.

The response current (I) can be obtained from Eq. 6.1, Assuming C_{dl} is constant for EDLCs, according to the following equation:

$$I = \frac{dQ}{dt} = C_{dl}\frac{dV}{dt} \tag{6.2}$$

where, t corresponds to the charge time.

If the variation of the applied voltage V is linear with time t, that is, $V = V_0 + vt$ (where V_0 is the starting voltage and v is the scan rate ($V\ s^{-1}$ or $mV\ s^{-1}$)), the relationship is defined as:

$$I = C_{dl} v \tag{6.3}$$

The current response is linearly dependent on the sweep rate as has been shown in Eq. 6.3. This results a clearly defined rectangular voltage (V)-current (I) cyclic voltammograms for different sweep rates (Figure 6.2(a-b)).

Not all the materials store charge through this process. Some materials like RuO_2, MnO_2, and so on store charge by the means of largely reversible Faradaic reaction with the electrolyte (Liu et al. 2018). In other words, the term "pseudocapacitance" is implemented to indicate the electrode materials (RuO_2, MnO_2) which have the electrochemical signature associated with a capacitive electrode (such as observed in case of the activated carbon), i.e., showing a linear dependence of the stored charge on the working potential, where charge storage is a contribution from various reaction mechanisms. The CV and GCD of these materials are not perfectly rectangular and triangular, respectively, as seen in case of the porous carbon materials, but they possess a little deviation from that, as can be seen in Figures 6.2(c-d). However, due to the involvement of Faradaic reaction mechanism, battery materials are often mistaken as pseudocapacitor materials, as shown in Figure 6.2(e-f) (Gulzar et al. 2016).

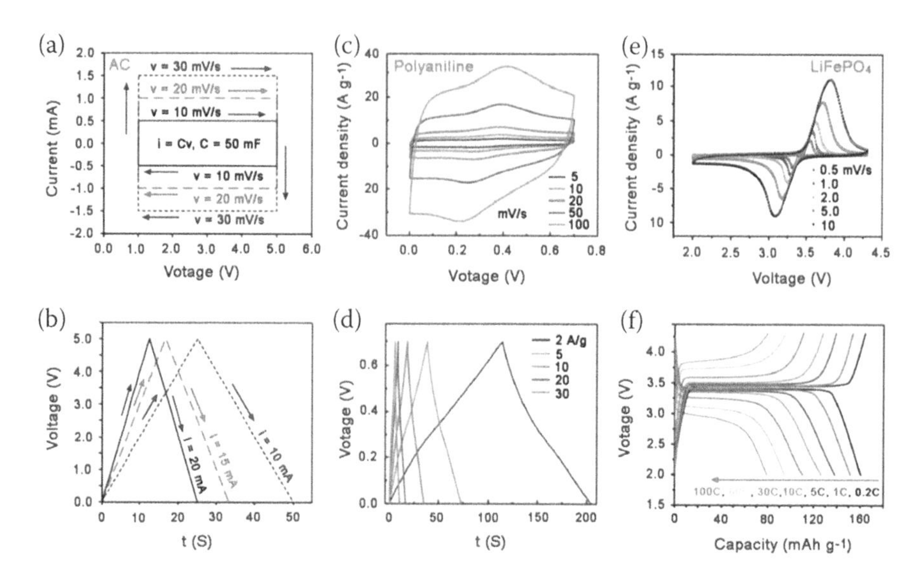

FIGURE 6.2 (a-b) Typical CV and GCD plots obtained for the supercapacitor type electrodes. (c-d) Nature of CV and GCD curves for pseudocapacitive type electrode material. (e-f) Nature of the CV and the GCD plots for battery-type electrode material.

Source: Reprinted with permission from (Liu et al. 2018). Copyright © 2018 John Wiley & Sons.

One major difference between the battery-type materials and the pseudocapacitor-type materials is that they originate from the Faradaic charge-storage mechanism comprising transfer of charge across the double layer, much like what happens in case of battery charging or discharging. But in this case, capacitance is associated with a special relation resulting from a thermodynamic cause between the limit of acceptance of charge (Δq) and the variation in the potential (ΔV); the result is that the derivative dq/dV or $d(\Delta q)/d(\Delta V)$ expressing the capacitance may be experimentally measured and formulated through dc, ac, or transient techniques (Brousse, Bélanger, and Long 2015).

Alternatively, at the state when the capacitor is discharged or charged with a constant current, the potential gets enhanced (charging) or reduced (discharging) at a rate that is again constant, as shown in Eq. 6.3. Hence, the charge-discharge curve is expected to be a triangular one, as shown in Figure 6.2(b). Over the past few years, notable progress has been achieved in the basic designing and understanding of electrode materials for energy-storage applications. Materials based on carbon, including CNTs (Senokos et al. 2016), ACs (Fuertes and Sevilla 2015), and graphene (X. Yang et al. 2013) exhibit as EDLC behavior, when the specific capacitance is influenced by the pore-size distribution and the pore structure (Raymundo-Piñero et al. 2006), number of carbon layers (Ji et al. 2014), and electrode surface area and surface state (Oh et al. 2014). To date, numerous measures have been taken to augment the EDLCs energy density, in the context of drawing a similarity between the carbon pore sizes and the electrolyte ion size

(J. Chmiola, G. Yushin, Y. Gogotsi, C. Portet, P. Simon, n.d.) tailoring the oxygen content, (Augustyn, Simon, and Dunn 2014) oxygen functionalizing the carbon surface (Tan et al. 2018) or adapting carbon having heteroatom (N, S, F, etc.) doping (Gueon and Moon 2015), as well as co-doping (S. Zhang et al. 2015), designing ionic liquids with a wide temperature range and high working voltage and (R. Lin et al. 2011) adopting redox-active species-based electrolytes (Fic, Frackowiak, and Béguin 2012). Nevertheless, supercapacitor based in the EDL mechanism for charge storage is unable to meet the sheltered necessities for devices with large energy density attributed to the inherited drawbacks, restricting their scaled up application.

6.2.3 Transition From Electrophysical Storage to Pseudocapacitive Storage

Pseudocapacitance originates from the energy storage that is Faradaic, arriving as a result of rapid redox reaction occuring on the surface or near the region of the electrode surface, where electrodesorption/electrosorption occurs as a result of charge transfer in the absence of any bulk phase transformation due to charging/discharging (Figure 6.2(b)). (The 1993) charge state (q) is dependent on the electrode potential corresponding to the Faradaic charge/discharge (Q) passed (B. E. Conway 1991a). The alteration in Q in comparison with the potential is denoted by the derivative, dQ/dV, which is associated with pseudocapacitance (C_p) (B. E. Conway and Pell 2003). Unlike EDL capacitance, which has the context of the potential-dependent accumulation of electrostatic charge (Figure 6.2(a) and (b)), pseudocapacitance is Faradaic in nature (Figure 6.2(c) and (d)) (B. E. Conway 1991a). It is noteworthy that the pseudocapacitance is somewhat different from the ideal Nernstian process involved in battery-type materials in which the Faradaic reaction happens at a constant potential (Figure 6.2(e) and (f)) (T. Brousse, Bélanger, and Long 2015). Conway et al. classified pseudocapacitance into three types: (i) surface redox system (2D), (ii) underpotential deposition (UPD) (2D), and (iii) intercalation system (quasi-2D) (Figure 6.3) (B. E. Conway 1991a).

As metal is charged with a potential, there occurs the creation of an monolayer adsorbed existing on the surface of electrode as a result of reduction of other metal ion, causing a less negative potential as compared to equilibrium potential, which is termed as UPD. (The 1993) deposition of Pb on Au is an UPD. (The 1993) UPD is applicable to metal deposition as well as other adsorbed layers, for instance, H from H_3O^+ or H_2O pseudocapacitance (Burke and Lyons 1986). The redox system is a typical instance of pseudocapacitance, related with the mechanism involving electroactive ion adsorption on the electrode materials' surface or in the vicinity of the surface and occurrence of charge transfer due to Faradaic reactions. Pseudocapacitance of the transition metal oxides originates as a result of the fast redox reactions occurring due to intercalation of protons (H^+) or alkali metal cations ($C^+ = Na^+, K^+$, etc.), as described below (J. P. Zheng and Jow 1995):

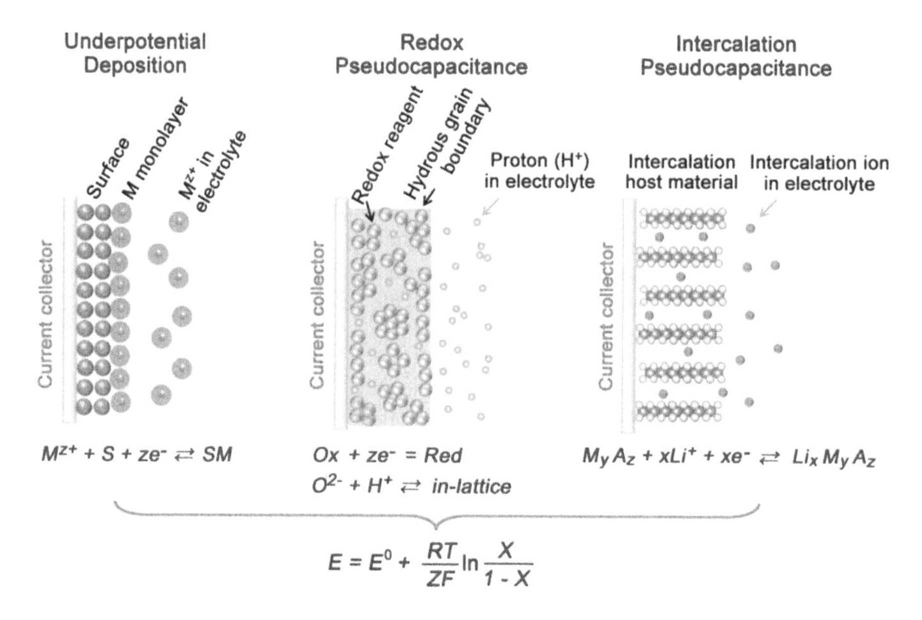

FIGURE 6.3 Various Faradaic charge storage mechanism. X represents the fraction of site occupancy related to the underpotential deposition $[O_X]/([O_X] + [Red])$ associated with the redox systems and the portion of occupancy related to the layer lattice corresponding to the systems of intercalation, respectively.

Source: Reproduced with permission (Becker 1957b). Copyright 2016, The Springer Nature.

$$MO_2 + C^+ + e^- \leftrightarrow MOOH \quad (6.4)$$

$$MO_2 + C^+ + e^- \leftrightarrow MOOC \quad (6.5)$$

Due to the process of charging/discharging, there is no chemical transformation taking place; however, a reversible functionalized molecular layer forms on the electrode surface due to Faradaic reactions. Potential of the electrode exhibit a linear dependence on the charge, which depends on the surface area of the electrode enclosed by electroactive ions. These properties are completely dissimilar as compared to the redox reactions occurring in a battery-type electrode.

Electrically conducting polymers (ECP) are able to release and store charge by the means of redox processes related to the polymer chains, which are π-conjugated during electrochemical doping–undoping, as has been represented by the following reaction (Doblhofer and Zhong 1991; Béguin et al. 2014):

$$[ECP] + nX^- \leftrightarrow [(ECP)^{n+}nX^-] + ne^- \quad (6.6)$$

At the time of oxidization (p-doping), the anionic species X^- from the electrolyte are incorporated into the backbone of the polymer and during reduction these are returned back into the electrolyte. The stripping and embedding of the counter ions lead to large specific capacitance values, matching with an electrochemical reaction of battery-type. However, ECPs are challenged by volumetric alteration during the charging and discharging, leading to poor cycling performance and poor mechanical properties of these brittle materials. (Béguin et al. 2014) So, various efforts have been put to refrain from these drawbacks. The most apt approach is to integrate carbon materials, e.g., Carbon Black (Peng Liu, Wang, and Wang 2014), Carbon Nano Tubes (Qin Yang, Pang, and Yung 2014), or graphene (K. Zhang et al. 2010)) with ECPs to improve the mechanical properties of the ECP. Intercalation pseudocapacitance is another type of Faradaic method taking place as a result of absence of a crystallographic phase alteration and taking place as a result of the intercalation adsorption of the quasi-2D electroactive species. It is dissimilar to the process of intercalation persisting in a battery, accompanied by phase transformation that is crystallographic at the time of the charge-transfer processes. Intercalation systems related to the pseudocapacitors come in association with the intercalation of Li^+ ions in the hosts, such as V_6O_{13} (H. M. Zeng et al. 2009), TiS_2 (G. Sun et al. 2014), MoS_2 (H. D. Yoo et al. 2016), and incorporation of H into Pd-Ag alloys and pristine Pd (J. Liu et al. 2018). Currently, novel 2D materials have been designed. Among them, transition metal carbides (MXenes) are recognized as distinct host materials as intercalation pseudocapacitors (Lukatskaya et al. 2013). High-volume pseudocapacitors have been stated to be synthesized with the help of intercalation of ions including Li^+, Na^+, K^+, NH_4^+, or Al^{3+} within the MXene layers. As has been denoted in Eq. 6.7, $Ti_3C_2T_n$, which is a typical mXene material, exhibits large volumetric capacitance as a result of change in the Ti oxidation state at the time of intercalation/ de-intercalation processes

$$Ti_3C_2O_x\ OH_yF_{2-x-y} + \delta e^- + \delta H^+ \leftrightarrow Ti_3C_2O_{x-\delta}OH_{y+}\delta\ F_{2-x-y} \tag{6.7}$$

These three mechanisms related to the pseudocapacitance are grounded on various Faradaic processes and take place in various kinds of materials; however, they bring similar features in context with the thermodynamic phenomenon. That is, a logarithmic relationship exists in between the extent of charge/discharge and the electrode potential, as shown in Figure 6.4 [9,18,24,53]:

$$E = E^0 \frac{RT}{nF} ln \frac{X}{1-X} \tag{6.8}$$

In the above equation, T denotes the temperature (K), E denotes the is the electrode potential (V), n denotes the number of electrons, R denotes the ideal gas constant (8.314 J mol^{-1} K^{-1}), X is the fraction of occupancy in the surface or lattice layer, and F is Faraday's constant (96 485 C mol^{-1}). In the case of sorption, which is electrochemical in nature, it is associated with the electroactive species taken place in accordance with electrochemical Langmuir isotherm (B. E. Conway 1995). As demonstrated in Equation (6.9), the pseudocapacitance can be defined as being

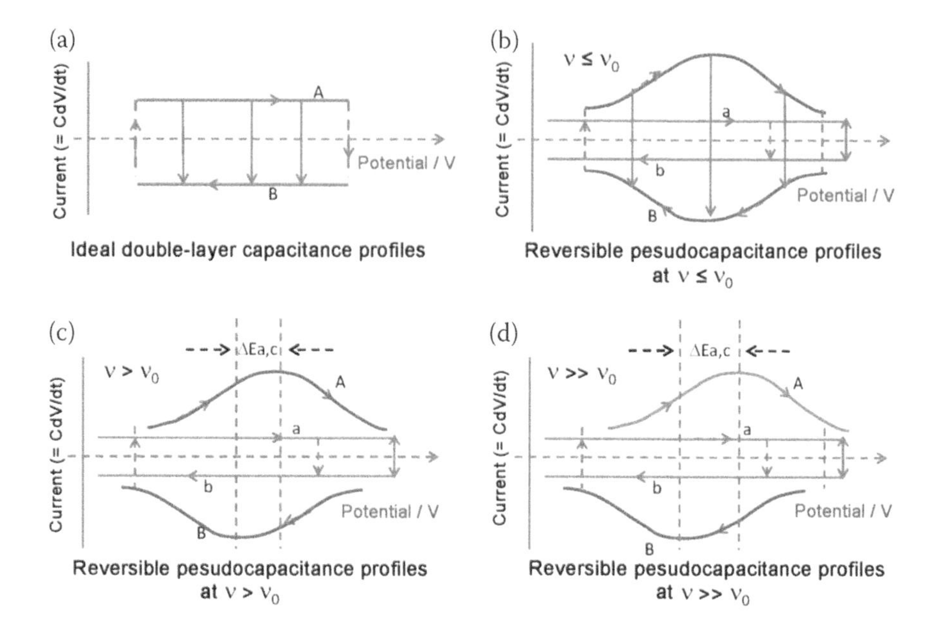

FIGURE 6.4 Cyclic voltammetry profiles of a) perfect double-layer capacitor and b–d) typical pseudocapacitors with varying sweep rates *v*.

Source: Reproduced with permission. (Simon, Gogotsi, and Dunn 2014) Copyright 1991, the Electrochemical Society.

denoted by Equation (6.10) (B. E. Conway 1991b), where q^* is the charge required for completion of monolayer sorption:

$$\frac{X}{1-X} = k\exp\left(\frac{VF}{RT}\right) \tag{6.9}$$

$$C_\phi = q_* \frac{dX}{dV} \tag{6.10}$$

The equations (6.9) and (6.10) imply that the pseudocapacitance C_ϕ is not a variable and exhibit the largest value at $X = 0.5$. The energy storage in the pseudocapacitors is quite similar to that of the energy-storage energy in EDL capacitors. However, the significant difference relies in the involvement of Faradaic charge transform in case of the pseudocapacitance. Particularly, for the pseudocapacitance, the potential of the electrode is dependent on the electroactive material conversion. For the maximum battery-type electrodes, a special electrode potential is denoted using the Gibbs free energies corresponding to the pure, well-defined 3D phases and also the concentration and/or composition of the solution ($\Delta G\text{-}nFE^0$) (Winter and Brodd 2004). Additionally, pseudocapacitors exhibit larger rate capability values as compared to the batteries attributed to the surface/near surface reaction.

6.3 ENERGY DENSITY AND POWER DENSITY

Energy density and power densities are two major parameters determining the performance of a supercapacitor. The total amount of charge that can be stored in a supercapacitor is expressed as the energy density of the supercapacitor, whereas power density is the measure of how quickly the energy can be dissipated. For the supercapacitors, the specific energy density is generally expressed in Wh kg^{-1} whereas the power density is expressed in W kg^{-1}. The energy density and power densities are measured of a device or in a two electrode system. The energy density can be obtained from the discharging specific capacitance according to the equation (J. G. Wang, Kang, and Wei 2015):

$$E_d = \frac{1}{2} C V^2 \tag{6.11}$$

where C is the specific capacitance of the device or the whole system, and V is the potential window.

The power density can be calculated from the energy density by dividing it by the discharging time. The equation for power density is (J. G. Wang, Kang, and Wei 2015):

$$P_d = E/t \tag{6.12}$$

where, t is the discharging time.

The plot of energy density against the power density at various current densities in log scale is termed as the Ragone plot. The energy density is directly proportional to the specific capacitance, and a popular method to increase the energy density is to increase the specific capacitance of the device. However, the working potential is squared to get the energy density. So, increasing the working potential of the device or the system is more effective in increasing the energy density. Incorporating pseudocapacitive element in the electrode material results in the increase in the energy density as redox charge-storage mechanism is able to storage more charge than the EDL system. However, even though the energy density increases, it decreases the power density as there is involvement of redox charge storage, which takes longer for interaction as compared to the surface-ridden charge storage. The power density of a device can be increased by using highly porous material so that large surface area is available for the charge storage.

Supercapacitor delivers lesser energy density as compared to those of batteries, whereas the power density is much higher as compared to the rechargeable batteries (Figure 6.5). However, the power density is less for supercapacitors as compared to the conventional capacitors, and the energy density is far more as compared to the conventional capacitors. In terms of energy and power densities, supercapacitors act as a bridge between the capacitor with high power density and the rechargeable batteries with large energy density.

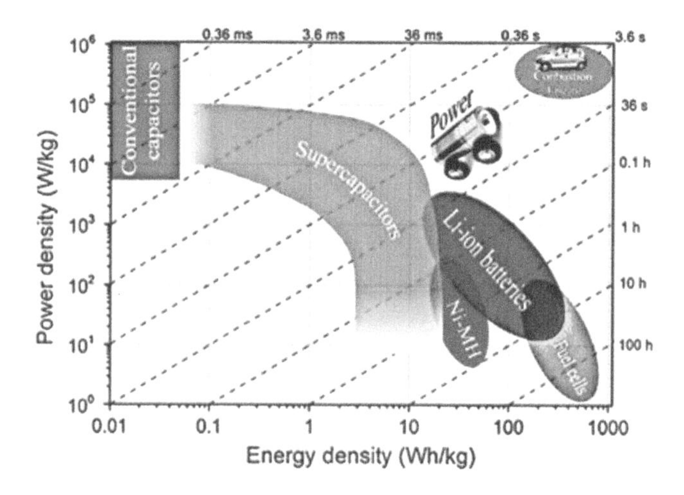

FIGURE 6.5 Ragone plots of different energy storage systems. (J. G. Wang, Kang, and Wei 2015).

6.4 ULTRA SMALL PLANAR DEVICES

The first generation of micro-supercapacitors was inspired by the planar 2D design associated with the thin film microbatteries. This facile design consists of a solid electrolyte and two electrodes of thin film (thickness <10 μm), which are placed on top of each other creating a complete supercapacitor cell (Figure 6.6(a)). For facile cost-effective production in bulk, this simple 2D design is preferred over the other designs. Nevertheless, from the viewpoint of the application, the planar 2D electrodes proves to be limited from the context of the amount of energy these are capable of storing in a small device. For this type of architecture associated with the device, preparing the electrodes more thick with the purpose of storage of augmented energy is not a feasible option as thicker electrodes are challenged by the electron and ion transport limitations, reducing the power density. Supercapacitor electrodes based on thin films are prepared and implemented using several methods; however as will be discussed in the following section, manufacturing methods, which will result in binder-free, conductive, and porous films, in turn lead to larger power and energy densities.

A change in the 2D structure is the one where both the electrode in the device will comprise numerous microelectrodes (fingers) interdigitally arranged on a substrate (in-plane interdigital electrodes), as illustrated by Figure 6.6(b). The electrode construction typically accompanied by the fabrication of thin film approaches together with the inclusion of a step involved in patterning of electrode before and after electrode deposition. This step of patterning mostly follows standard microfabrication methods, with a few exceptions. The in-plane design associated with the interdigitated electrodes has a few advantages over the traditional 2D structure. Presence of the electrodes in the similar plane will result in the integration and fabrication of micro-supercapacitors. Also, an important cutoff in the ion transport resistance is possible to be achieved by tuning the gap between the two

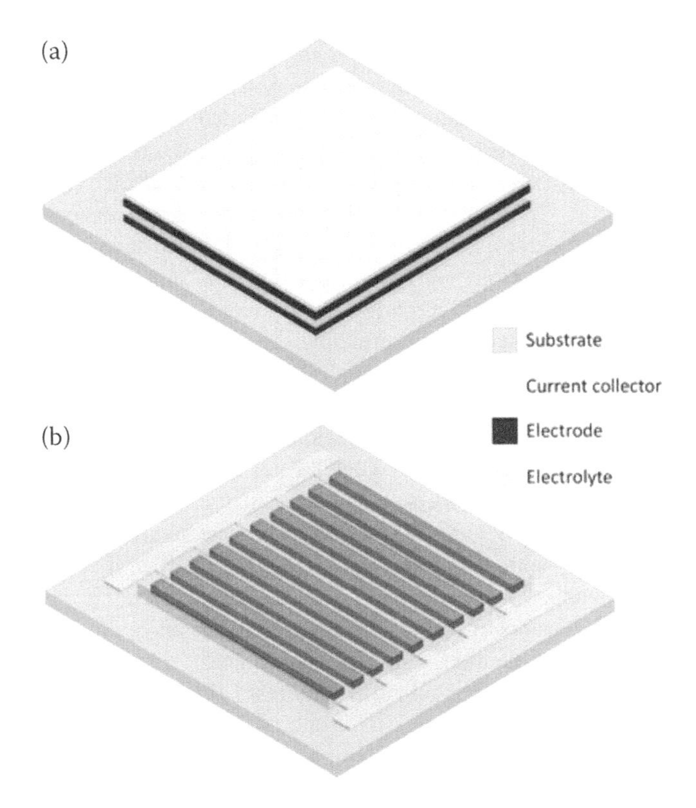

FIGURE 6.6 Micro-supercapacitors grown on-chip with (a) traditional architecture of 2D and (b) interdigital in-plane electrode architecture (Beidaghi and Gogotsi 2014).

electrodes, applying high-tech microfabrication techniques. Yet another positive side related to this device architecture is the improved accessibility associated with the electrodes as the edges associated with the microelectrodes are also exposed to the electrolyte. This is especially significant for the electrodes fabricated with the help of the layered materials, including graphene, because arranging electrodes one after the other side-by-side improves the access of ions amid the electrode material layers, leading to augmentation in the rate capability. However, this side-by-side arrangement of the electrode could cause reduced areal energy density in combination with highly thin electrodes as compared to the conventional 2D design as bigger footprint area is necessary for the device to contain same quantity of the electrode material. However, in several cases, the benefit related with the in-plane architecture overshadow this disadvantage.

6.5 DEVICE DESIGN PARAMETERS

Flexible supercapacitors are perpetually taking the place as desired energy-storage platform for driving devices credited to the transportable applications, such as diagnostics based on point-of-care and Internet-of-Things (IoT). A big advantage in comparison with the supercapacitors is their competence to combine with the sporadic energy sources made available by energy harvesters (R. Song et al. 2015). In this way, flexible supercapacitors have been intended to bridge the gap between batteries and energy harvesters. So far, research based on the flexible supercapacitors was mainly concerned with the growth of novel electrode and solid-electrolyte materials. Graphene, conducting polymers, carbon nanotubes, various 2D materials, metal oxides, and biomass-derived carbons have been studied as electrode materials in flexible supercapacitors (Dong et al. 2016). Current developments are challenged by configurational arrangement concerned with the flexible components clearly because different principles of design importantly influence the formation of the double layer (Shao et al. 2018). The experimentally observed augmented capacitance value related with micropores is decisively credited to the desolvation, which is partial of the ions belonging to the electrical double layer. This phenomenon was then supported by theoretical methodical insights meant to stem correlations between pore size and capacitance for versatile pore regimes (J. Huang, Sumpter, and Meunier 2008). However, one significant challenge considered when dealing with micropores was the EDL formation reversibility, upsetting the cyclic stability of the device. This delivers prospects for the preparation of electrode surfaces nanostructured in nature to reach an optimization between longer cyclic life in combination with high capacitance.

Attendant with thickening scientific interest, this field of research has seen developments that are technical and engrossed on the designing of supercapacitor energy-storage devices for driving mobile electronics, including flexible sensors for essential implantable medical diagnostics, physiological parameters, power T-shirts, smart devices, and electronic textiles (Figure 6.7) (Mondal and Subramaniam 2019).

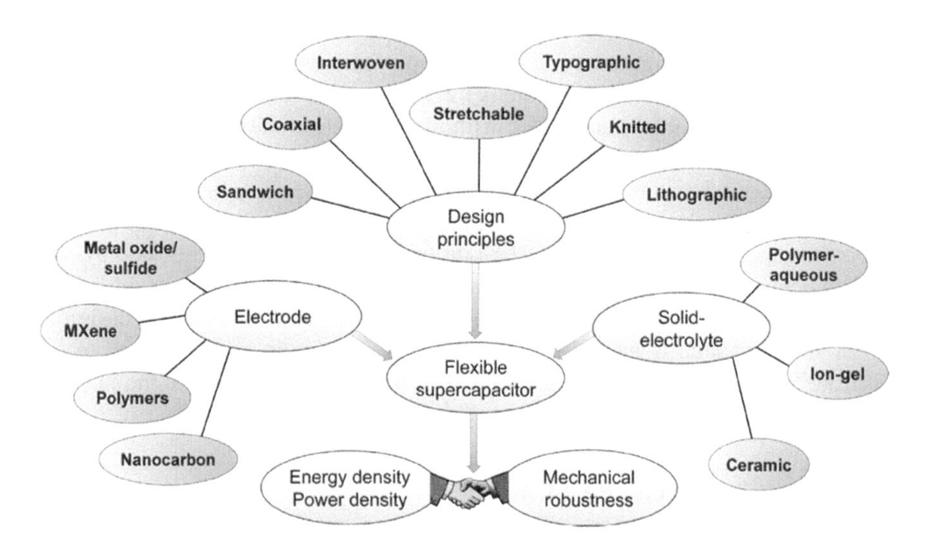

FIGURE 6.7 Track toward understanding supercapacitor energy storage devices meant for wearable applications.

Here it is intended to outline several aspects of principles of design concerned with the growth of supercapacitors which is solid-state in nature. A quantitative and qualitative comparison of various kinds of design methods, based on the technical feasibility and the exhibited properties, have been presented. The energy density delivered by the wearable supercapacitors is straightaway related to the accessible surface area, electrical conductivity, and pore structure associated with the electrode material in plus possessing mechanical robustness. The primary electrode material, which has proved to be capable of enhancing the energy density, comprises transition metal sulfides and oxides along with their composites with carbon. Such compounds (MnO_2, RuO_2, IrO_2, graphene-MnO_2, FeS_2, CNT-MnO_2, Co_2S_3, Cu_2S, and Ni_3S_2) show rich electrochemical redox activity credited to existence of unpaired valence shell electrons and the multiple oxidation states. In this context, the grouping of spinel/mixed oxide Zn_2SnO_4 (ZTO)-MnO_2 on carbon microfibers exhibited augmented capacitance of the device (Dong et al. 2016). An alike enhancement in the performance together with thermal stability has also been seen while implementing electrodes based on transition metal sulfides. Carbon composites, unspoiled-conducting polymers together with their doping variations, have been extensively investigated as electrode material, specifically because of their high electrical conductivity, mechanical stubbornness, and simple solution processability that results in seamless immobilization upon the flexible substrates implemented in textiles, as shown in Figure 6.8 (Dong et al. 2016).

An important difference has been centered on the growth of carbon material that is nanostructured as electrodes (Figure 6.8). Strongly unified Van der Waals force is one general problem encountered when handling the nanocarbon materials, resulting in their restacking and hence limiting processability in solution-state. Graphene electrodes, which are interdigitated, or laser scribed upon flexible supports, show high capacitance. Moreover, the occurrence of abundant elemental doping paves a path for attaining larger energy density (Dong et al. 2016). Large volumetric capacitance is attained applying self-supporting free standing mXene films. The complexity of processability in solution state of mXenes offers flexibility in design methods for fabrication of electrode, including patterned interdigitation and spray painting.

Equally crucial for enhancing the flexible supercapacitors performance is the introduction of a solid electrolyte. The novel research in the case of electrolytes in solid form are thus mainly absorbed on ion gels that reaches up to 4 V, thus creating a probability of mixing with wireless communication devices and transmitters. For example, Ahn et al. has reported an ion-gel electrolyte-based flexible supercapacitor operable at 4 V. A ceramic-based solid electrolyte Li_2S P_2S_5 electrolyte based on glass ceramic have been reported to implement as a flexible supercapacitor along with implementing multiwalled carbon nanotube electrodes (Dong et al. 2016).

Remarkable novelties in the electrode and electrolyte designing have transferred the focus on the importance of various approaches of design to combine the electrolyte and electrode to create a functional interface. The nature of the electrochemical interface is mainly dependent on the choice of the electrodes and the applied designing technique. In such a way, identical electrode materials have

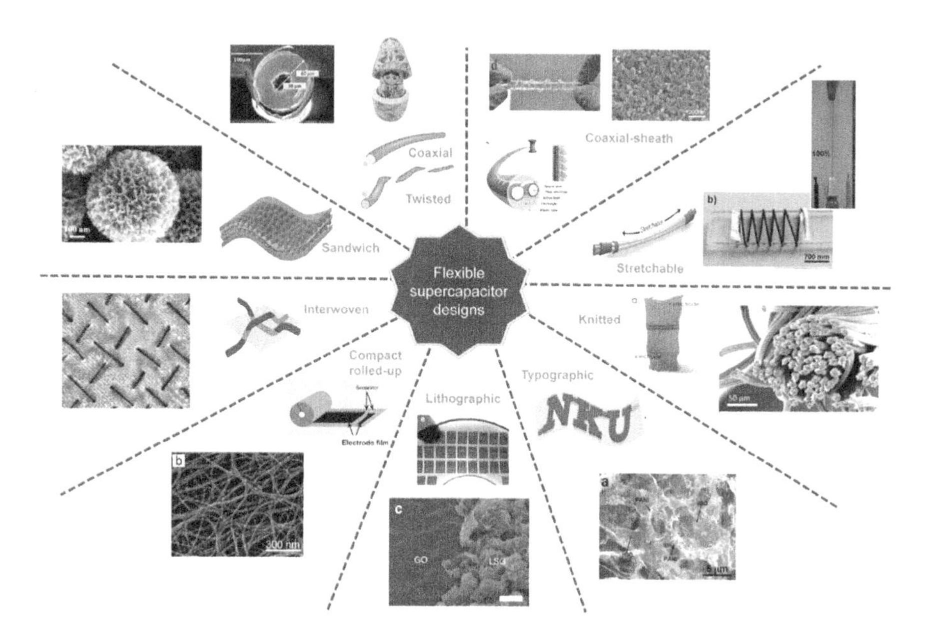

FIGURE 6.8 Illustrative schematics and electron microscope images.

Source: ((Jha, Hata, and Subramaniam 2019) and (Jha, Babu, et al. 2020), (T. Qin et al. 2017), (Niu et al. 2011), (Zhibin Yang et al. 2013), (El-Kady and Kaner 2013), (Harrison et al. 2013), (Jha, Ball, et al. 2020)) Images are adapted with permission from refs ((Jha, Hata, and Subramaniam 2019) and (Jha, Babu, et al. 2020), (T. Qin et al. 2017), (Niu et al. 2011), (Zhibin Yang et al. 2013), (El-Kady and Kaner 2013), (Harrison et al. 2013), (Jha, Ball, et al. 2020)). Copyright 2019 American Chemical Society, Copyright 2020 American Chemical Society, Copyright 2017 Wiley, Copyright 2011 RSC, Copyright 2013 Wiley, Copyright 2018 Wiley, Copyright 2013 Nature Publishing Group, Copyright 2013 RSC, Copyright 2020 American Chemical Society.

revealed to create various electrode–electrolyte interfaces as a result of various design principles (J. G. Wang, Kang, and Wei 2015). For example, electrodes on the basis of CNT film have been set up to act as minute relaxation time constants in lithographically fabricated compact devices (Laszczyk et al. 2015). Nevertheless, identical CNTs have been reported to be implemented in a washable and wearable device as an interwoven wire (Jha, Hata, and Subramaniam 2019). Consequently, the electrochemical interface and the design of fabrication are interdependent. Henceforth, the pursuit for a perfect interface is inadequate with the absence of choice of the design principle and electrode material.

Several reports are available on fundamental detailed theoretical understanding of interface having microporous structures (J. Huang, Sumpter, and Meunier 2008). Fascinatingly, the designs which are textile-based such as interweaving and knitting causes transverse and longitudinal charge polarization, causing a unique interfacial structure (Jha, Hata, and Subramaniam 2019). A call-off between power and energy density in these devices is solved to some extent by designing interdigitated co-planar electrodes (Figure 6.9(b)). As the design approaches related to lithography

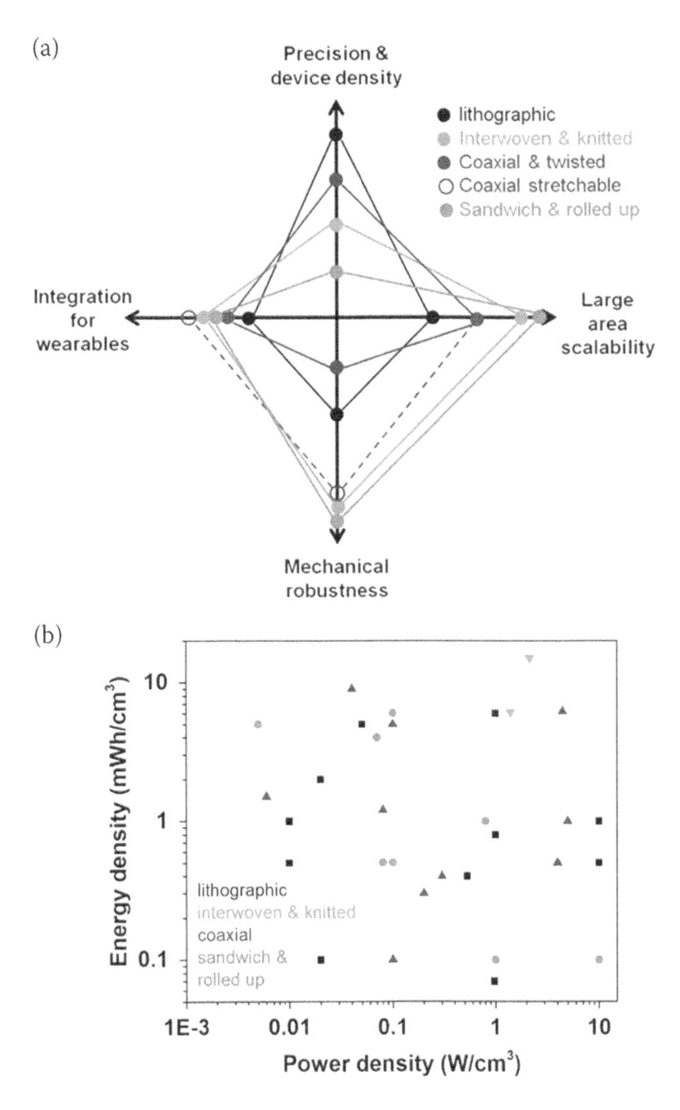

FIGURE 6.9 (a) Spider plot drawing a qualitative comparison between the of the merits of various methods in fabricating supercapacitors for wearable purposes. (b) Quantitative comparison of the flexible supercapacitor performance implementing different design approaches (Forouzandeh, Kumaravel, and Pillai 2020).

and spray coating are well established in case of the coplanar devices, additive manufacturing, typography, and printing require further developments having larger possibility for scalable implementation (Figure 6.9) (Shao et al. 2018). Various devices implementing threads and yarns with various functional groups have been designed by interweaving, knitting, additive manufacturing, and coaxial sheathing routes (Shao et al. 2018). Coaxial stretchable cable-based supercapacitors have been attempted by applying elastic substrates and carbon–polymer composite-based twisted yarns. A step forward in thrifty fabrication methods has been attained by employing micro-supercapacitive junctions (sewcaps) with CNT wires (Jha, Hata, and Subramaniam 2019; Jha, Jain, and Subramaniam 2020). Interesting designs within fabrication of electrodes include the Japanese paper-art forms including kirigami and origami (Dong et al. 2016). A comprehensive summary of several design principles in flexible supercapacitors has been discussed by Kaner and co-workers (Dong et al. 2016).

Therefore, this chapter attempted to illustrate the comparison of the various strategies in accordance with two separate parameters: (a) a qualitative comparision based on setting, such as packing density of the device, integration simplicity into platforms that are wearable, mechanical robustness, and scalability, which decides its technical viability (Figure 6.9(a) and (b)), a comparison which is quantitative based on the power density and energy density, as illustrated in the Ragone plot (Figure 6.9(b)). In addition to the scalability, a comprehensive assessment is also possible to be drawn classifying wearability, precision, density of the device, and mechanical strength as the other parameters. Furthermore, a quantifiable descriptor of performance has been illustrated by employing a Ragone plot (Figure 6.9(b)), which plots the power density and energy density of devices made with various principles of design. The insights have been represented in Figure 6.9 and have clearly been selected from relevant literature reports (Dong et al. 2016).

7 Self-Powered Supercapacitor

7.1 MECHANISM

As a result of the tremendously growing consumption of energy in modern societies, the growth of viable and green energy sources has emerged as one of the most significant research fields. Sources of green and sustainable energy, including wind and solar, are intermittent because the harvesting of energy can be disconnected in the absence of sun or or wind (Simon and Gogotsi 2008; B. Tian et al. 2007). Other energy-harvesting technologies, including nanogenerators, can efficiently transform the vibration energy from the living environment to electrical energy (Z. L. Wang and Song 2006), but these are not appropriate for continuously driving small electronics attributed to their limited power output (G. Zhu et al. 2010). Hence, this makes association of the energy-storage systems with the energy-harvesting devices a mandate to provide balanced and lasting energy that can be regulated. Supercapacitors and batteries belong to the advanced electrical energy storage (Simon and Gogotsi 2008). Supercapacitors are exclusive to batteries in terms of the power, green benignancy, cyclic stability, and safety (Channu et al. 2011). Portable/wearable personal electronics are applicable to electronic skin, wearable displays, and distributed sensors (Yuan et al. 2012). Self-powered networks of sensors are expected to play major roles in the upcoming years in controlling the world economy (Z. L. Wang 2010). The "self-powered nanotechnology" was able to ensure a network of sensors to operate sustainably and independently in the absence of any battery or by minimally expanding the lifetime of a battery (Xiao et al. 2011). Ultra-small self-powered nano systems exhibiting multi-functionality, as well as low energy consumption, are possible to develop by implementing nanomaterials and nanofabrication technologies (S. Xu, Hansen, and Wang 2010).

For eliminating energy crisis, researchers have established novel methods for integrating two devices (energy storage and harvester) through intrinsic or extrinsic means to attain a self-driven system that will be suitable for different implementations in the regime of microscale to macroscale (H. Wei et al. 2017). An integration of a nanogenerator, a supercapacitor with a fuel cell and a solar cell through extrinsic and/or intrinsic way has garnered much attention of various research groups (T. Li et al. 2016). Herein, as compared to the external integration, the intrinsic integration approach is advantageous as it needs a complicated power management requiring further manufacturing cost (X. Fu et al. 2018). Thus far, energy devices integrated intrinsically are efficient for accumulating, transforming, and saving electrical energy within a sole device; they have proven to be of tremendous interest for fundamental science experimentation and development of product attributed to their versatile device plan and ideas (Xue et al. 2012). Within

DOI: 10.1201/9781003174554-7

the several kinds of intrinsically consolidated energy devices developed as yet, a self-charging power cell (SCPC) implementing the concept of piezoelectric in integration with a battery has been reported by Prof. Z. L. Wang and co-workers (Xue et al. 2012); it projects itself to be of key interest. This is ascribed to the capability of this SCPC device to obtain electrical energy from mechanical motion by using a polymer separator piezoelectric in nature and store it in the battery-type electrode *via* a novel "piezoelectrochemical conversion process" (Xue et al. 2014). Subsequently, researchers established various actions, including separators consisting of PVDF polymer with enhanced Li-ion transfer channels for SCPCs. Recently, a solid electrolyte was fabricated by Xue et al. (H. He et al. 2017) having a porous PVDF separator to enhance the linked-energy shifting process in SCPCs. Similarly, SCPCs made by application of supercapacitor electrodes are recognized as self-powering supercapacitor cells or piezoelectric supercapacitors (Pazhamalai et al. 2018). The cells may have an ability of rapid charging as compared to the traditional cells due to the larger power density of the supercapacitor electrodes as compared to the battery-type electrode in the SCPCs. The device plan of SCSPC involves a piezo-polymer having gel-electrolyte coating sandwiched between supercapacitor electrodes (Pazhamalai et al. 2018). To date, electric double-layer capacitive electrodes, intercalative pseudocapacitance materials ($MoSe_2$), and pseudocapacitive electrode (MnO_2) have been used in reported SCSPC (Pazhamalai et al. 2018). However, the self-charging effectiveness associated with the SCSPC is low, limiting their applications practically; however, an explicit correlation among the role of the electrolytes and the electrodes on the performance of SCSPC is yet to be known. More significantly, the process of energy storage and conversion in an SCSPC, connected directly to the "piezoelectrochemical effect," is established only on the basis of theoretical prototypes and is yet to be addressed in scientific research (Pazhamalai et al. 2018). Therefore, coming up with strategies to realize the "piezoelectrochemical effect" is expected to provide novel mechanical to electrical energy storage and conversion process. This will prove beneficial for the optimization of the important factors for the betterment of energy-conversion efficiency (Krishnamoorthy et al. 2020).

7.2 STATE-OF-ART DESIGN AND APPLICATIONS

There can be several applications of the self-powered supercapacitor in the field of wearable smart electronics, flexible electronics, healthcare applications, etc. Similar to these applications, the state-of-the-art design of self-powered supercapacitor is discussed in the next sections.

7.2.1 Wearable Electronics

The wearable electronics devices include the wearable bio-medical and biosensor devices. As these electronics need electric power to function, portable electronic systems form an integral portion of portable devices. Actually, the energy-storage modes in these electronic devices should exhibit the properties like flexibility and comfort for the user. Some significant works are discussed,

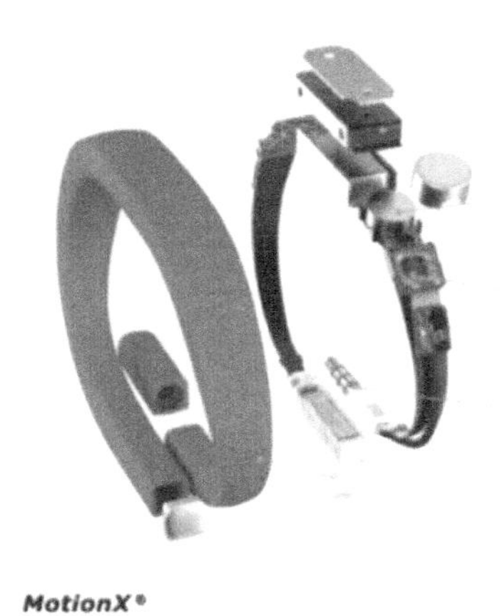

FIGURE 7.1 Application of wearable electronics.

which report works on devices designed for power conversion and storage applications implemented in mobile devices. The main focus is on the fabrication of piezoelectric generators, solar cells, supercapacitors, and batteries to be implemented in the portable device applications, as shown in Figure 7.1. These devices are generally implanted with fabric. These are restricted to devices made up of fiber and ribbon (Sodano et al. 2017).

7.2.2 Flexible Electronics

Solid-state supercapacitors are mainly associated with flexible applications by present and future generations. Flexible supercapacitors can be easily associated with wearable clothing and are suitable as a source of power for various electronics devices, including mobile phones. The power generated by piezoelectric harvesters are usually stored in large storage and implemented for charging mobile phones. For instance, there has been a report about a high-power T-shirt termed as a "sound charge," which is able to generate electricity under the pressure of sound waves. The T-shirt when tested could produce enough electricity to recharge two basic mobile phones through the whole weekend, as shown in Figure 7.2 (Ali, Albasha, and Al-Nashash 2011).

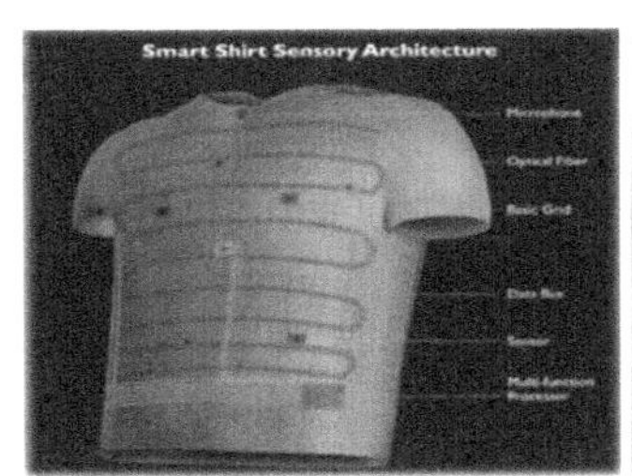

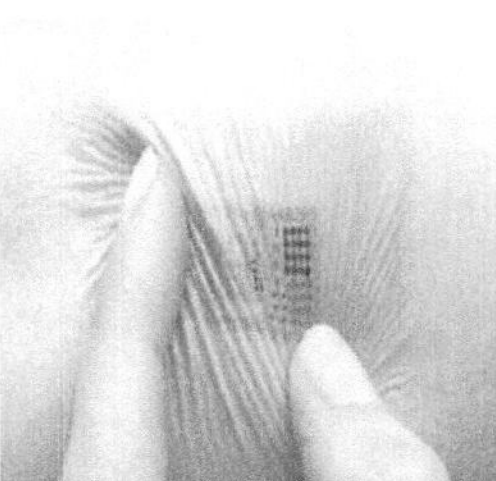
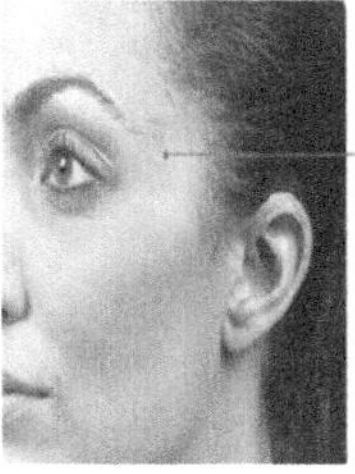
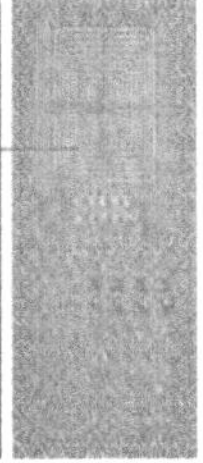

FIGURE 7.2 Application of flexible electronics.

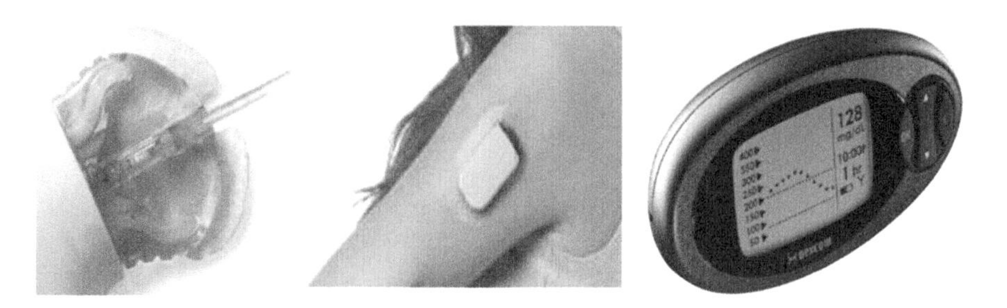

FIGURE 7.3 Continuous glucose monitoring (CGM) sensor, receiver and transmitter.

7.2.3 Healthcare Applications

Piezoelectric-charged supercapacitors are implemented in several combined health systems where the power requirement can shoot up to 1 million microwatts. These supercapacitors are applied for insulin pumps, pacemaker operations, and various other healthcare systems. Ongoing glucose-monitoring systems (CGM) are able to monitor blood sugar levels for the entire day. CGM users push in a small sensory wire below their skin to implement an automated device. This attachment possesses a CGM sensor enabling the sensor to measure glucose readings during the day and night. A reusable, small transmitter follows the sensor and sends real-time readings to the receiver offline, availing the data to the user. In some applications, a well-matched smart device incorporated with a CGM application plays the role of a display device. An efficient receiver or smart device displays the present sugar levels, as well as the past records of the sugar levels. A well-matched CGM receiver and smart device can be integrated to send alerts to the user when the glucose limit is reached (Ali, Albasha, and Al-Nashash 2011).

With the development of technology in the field of wireless network and microelectromechanical systems, smart sensors fabricated to be set up in remote locations, including medical sensors and pull-sensing health sensors implanted in the human body, have lost a CGM (continuous glucose monitor) device. It displays the "real time" glucose reading and leanings in glucose levels. The glucose level reads under the skin every 1–5 minutes (10–15 min delay). This shows that low and high glucose monitors are able to switch on the alarms and notify diabetes management practices, as shown in Figure 7.3 (Bharathi Sankar Ammaiyappan and Ramalingam 2021).

8 Design and Fabrication of Planar Supercapacitor Electrodes

8.1 FUNDAMENTALS OF ELECTROCHEMICAL INTERACTION

Micro supercapacitors can be categorized in various ways, based on the storage mechanism, the type of electrolyte implemented, and the kind of materials employed. Depending on the storage mechanism, MSCs can be classified into two categories. (1) Electric double-layer capacitors, where the capacitance mainly originates from the oppositely charged double-layer formed at the interface because of charge separation. At the time of charging, the electrode surfaces will be charged as negative and positive when a voltage is applied; consequently, they will attract positively charged and negatively charged ions. As the electric field is scattered, the mutual attraction existing in between the charges prevents the two layers from merging with each other and results in the formation of a stable electric double-layer. The energy is stored likewise through the formation of oppositely charged double layers. At the time of discharging, electrons are transported from negative electrode to positive electrode, discharging through load and producing a current; with this process, the energy is released (Xiaoyu Zhao et al. 2019). Materials based on carbon are typical EDLCs origin. Credited to the high surface area, these can contribute significant EDL storage. (2) Pseudo-supercapacitors, which depend on quick reversible redox reactions to give rise to capacitance. At the time of charge, ions will accumulate at the interface under applied potential; this step is followed by a redox reaction associated with electrode, and synchronously, this process leads to energy storage. At the time of discharging, the adsorbed ions are released from the electrode surface, resulting in dissipation of the stored energy through the external circuit. In general, as ions can get into the electrodes and take part in redox reactions, pseudo-supercapacitors are able to provide capacitance, which is 10–100 fold higher (Xiaoyu Zhao et al. 2019). Metal oxide and conjugated polymers are the pseudo materials, which can deliver high faradic reactions.

8.2 DESIGN DEPENDENT ENERGY-STORAGE MECHANISM

As a result of different methods implemented in designing the finger electrodes and the associated devices, the electrochemical performance the device exhibits can significantly vary. This result indicates that the method of preparation itself significantly influences the device properties. This design-dependent device performance may be because of the next motives:

DOI: 10.1201/9781003174554-8

For a particular type of materials, only little amount of variation in the electrode fabrication would cause change in the different types of microstructures, which will in turn affect the properties of the device starting from the ion transport ability of ions and electrons to the resistance offered by the electrode for charge transfer. Secondly, the electrochemical performance of the MSCs are influenced by both the broadness of the electrodes and also the gap between two neighboring electrodes. The preciseness of the above parameters will be also swung by the various methods of fabrication, resulting in difference of performance. Hence, here, a few significant fabrication methods previously reported resulting in patterned electrode finger arrays are outlined. In addition, various rewards and drawbacks of individually approach is compared and realized to offer guidance for manufacturing high performance MSCs.

8.2.1 Screen Printing

This approach implements ink to make specific patterns upon substrate by using a screen. It has the benefits of fast processing, robustness, and is relatively cheap; numerous substrates are available for direct stamping. Hence, this approach is very appropriate for the designing of MSCs, particularly in a large-scale production (Das et al. 2020). The conventional approach associated with the printing entails taking prints of the current collectors and active electrode materials separately. For instance, as shown in Figure 8.1(a), Chih et al. (Chih et al. 2019) demonstrated the synthesis of the all-screen-printable approach to produce flexible all-solid-state MSCs. A coat of current collector Ag was screen-printed on the layer of electrode. The device so fabricated exhibited an outstanding cyclic life corresponding to the capacitance retention of >99% over 15,000 cycles (Figure 8.1(b)). Furthermore, Liu et al. (Li Liu et al. 2018) successfully assembled a flexible MSC through initially screen-printing current collector Au and printing Ag@polypyrrole (Ag@ Ppy) ink later, as shown in Figure 8.1(c). After dumping the gel electrolyte, the fabricated devices showed an admirable electrochemical performance associated with large energy density corresponding to the value of 4.33×10^{-3} mWh cm^{-2} and large cyclic steadiness, together with great mechanical flexibility. Furthermore, through screen printing, it is also possible to print MXene materials. For instance, Yu et al. (Lianghao Yu et al. 2019) synthesized disheveled nanosheets of N_2-doped MXene (MXene-N), implementing the template melamine formaldehyde. Over the whole approach of optimization viscosity, the ink was adjusted to fabricate devices based on nanosheets of MXene-N by screen printing. The doping by N and wrinkled structures improved the redox activity and conductivity, and the MSCs fabricated could deliver an areal capacitance corresponding to the value of 70.1 mF cm^{-2} and excellent mechanical strength. In yet other case, obtaining the pieces of multilayered MXene and unelected cursor in the aqueous inks, Abdolhosseinzadeh et al. (Abdolhosseinzadeh et al. 2020) designed on paper MSC, as shown in Figure 8.1(d).

The resulting MSCs show a more areal capacitance (158 mF cm^{-2}). Printing enables large-batch production in a shorter period compared to other production methods. Furthermore, screen printing allows printing of both electrode materials and electrolytes in different steps to fabricate all-solid-state MSCs.

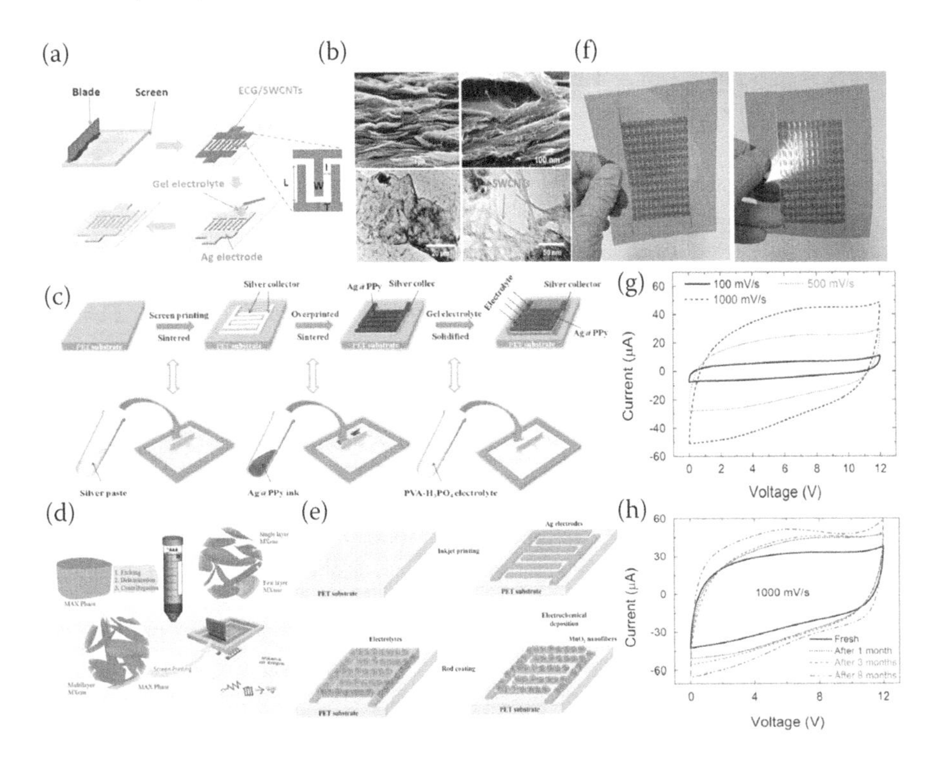

FIGURE 8.1 (a) Representation of all screen-printable devices. (b) SEM of hybrid electrode materials. (c) Construction procedure of the MSCs fabricated through ink printing. (d) Diagram for straight-away screen printing. (e) Fabrication of the flexible MSCs. (f) Photos of Kapton-based devices. Cyclic voltammetry profiles corresponding to MSC array at various scan rates and different periods.

Source: (a, b) Reproduced with permission (Chih et al. 2019) (Copyright © 2019, Royal Society of Chemistry). (c) Reproduced with permission (Li Liu et al. 2018) (Copyright © 2017, John Wiley and Sons). (d) Reproduced with permission (Abdolhosseinzadeh et al. 2020) (Copyright © 2020, John Wiley and Sons). (e) Reproduced with permission (T. Cheng et al. 2019) (Copyright © 2019, John Wiley and Sons). (f–h) Reproduced with permission (J. Li et al. 2017) (Copyright © 2017, American Chemical Society).

8.2.2 Inkjet Printing

This type printing is an approach that includes ink spraying on the surface of the substrates to produce different patterns. As compared to the additional technologies, it enjoys the rewards, including large precision, capacity of mass production, and feasibility of room-temperature manufacturing. Furthermore, through this process, it is possible to print directly on the substrate, which makes this approach very suitable to obtain thin films in energy-storage devices, solar cells, etc. (M. Singh et al. 2010). In the approach of the ink-jet printing process, the primary aim is to prepare an ink having a suitable flexibility. The most traditional way to ink configuration associated with the material is considered, and then its direct substrate printing to procure the devices. For example,

Wang et al. attempted a flexible MSC through printing an electrode-implementing 3D metal-ion-doped MnO_2 sheets. Credited to the metal ion incorporation, electrochemical properties of the designed device were able to improve significantly. Additionally, Pang et al. (Pang et al. 2015) made asymmetric MSCs having lamellar $K_2Co_3(P_2O_7)_2{\cdot}2H_2O$ nanomaterials and nanosheets of graphene as anodes and cathodes, respectively, through inkjet printing. The designed MSC showed a large volumetric capacitance of 6.0 $F.cm^{-3}$. Additionally, along with conventional ink-jet printing approach, using this method in combination with other methods, including deposition, also provides a suitable method. For example, Cheng et al. (T. Cheng et al. 2019) produced Ag electrodes, which were interdigitated in arrangement with the substrate based on PET through the process of inkjet printing. It was followed by procuring porous nanofiber-like electrode structures by the Ag electrodes coated by MnO_2 electrochemical deposition, as shown in Figure 8.1(e). Credited to the advantages coming in place with the structures, the fabricated MSC could cause high capacitance attaining 46.6 mF cm^{-2} with 86.8% capacitance retention over 1000 times of cycles of bending to 180°. It is noteworthy that ink-jet printing is applicable for the fabrication of large-scale MSC arrays in a facile and rapid way. For instance, Li et al. (J. Li et al. 2017) could effectively combine more than one device through the method of ink-jet printing, which eliminated the drawback associated with low voltage, which is generally encountered within a particular device; therefore, this becomes practical. Exclusively, the researchers designed MSCs based on graphene with the help of a simple full-inkjet printing technique. With the help of optimization of the thickness of the electrode material and different other parameters, the MSCs synthesized produced an areal capacitance of 0.7 mF cm^{-2}. More significantly, greater than hundred MSCs (Figure 8.1(f)) could be successfully linked to result in a power banks, in turn leading to effective combination of large-scale supercapacitor arrays. The assembled MSCs could be charged up to 12 V (Figure 8.1(g)), and the electrochemical property could be retained for more than 8 months (Figure 8.1(h)), confirming excellent integration and electrochemical properties of the MSCs. In a nutshell, even if the fabrication efficiency of printing is substandard as compared to the screen printing, the high resolution projects itself to be an important plus point. Furthermore, inkjet printing also makes it possible to select from a broad range of substrates and electrode materials, causing the integration of flexible MSCs with high performance. Moreover, large working potential and energy density is attainable through printing highly integrated MSCs on flexible substrates. This is significant for the practical applicability of MSCs in systems with large voltage. Inkjet printing can be further improved by precisely controlling the scope of the electrodes' size and in between spaces of the electrodes in combination with the enhanced ink quality, and hence, the production of high performance MSCs.

8.2.3 Photolithography

This approach uses patterns made out of photoresist, achieving high product determination, and resulting in several complex plane patterns. Thus, it is often

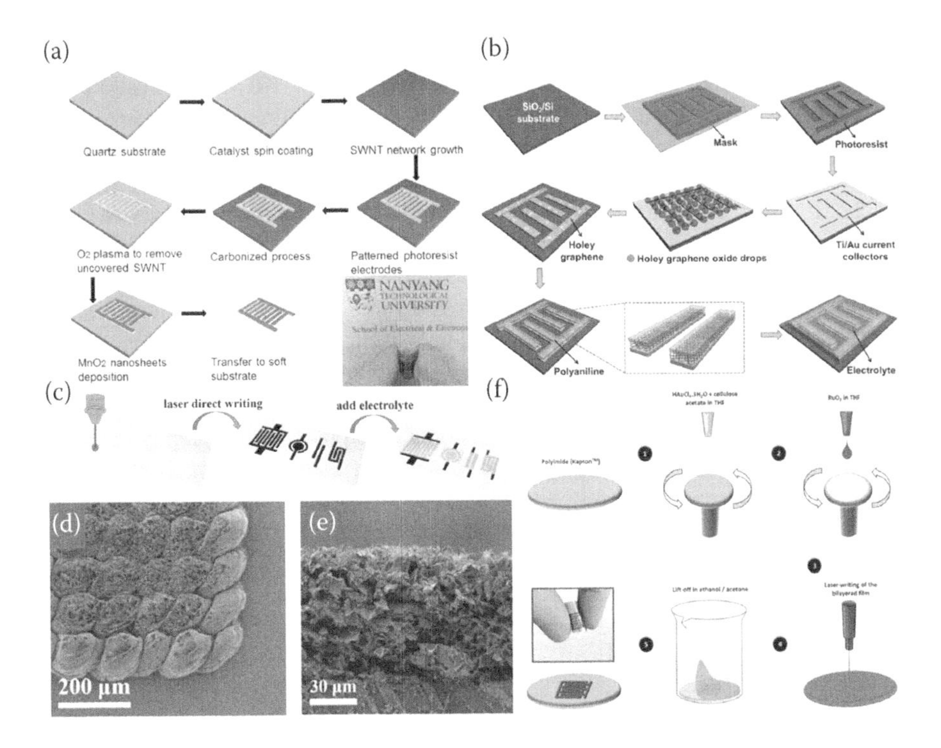

FIGURE 8.2 (a) Schematic illustration of soft substrate based MSC fabrication; (b) MSCs associated with heterostructures of holey polyaniline/graphene. (c) Designing of MSCs based on laser-induced graphene (LIG) films. (d) Top-view and (e) cross-sectional SEM images corresponding to LIG films. (f) Designing approach of RuO_2-based flexible MSCs.

Source: (a) Reproduced with permission (L. Sun et al. 2016) (Copyright © 2016, Elsevier). (b) Reproduced with permission (X. Tian et al. 2016) (Copyright © 2016, Springer Nature). (c–e) Reproduced with permission (X. Shi et al. 2019) (Copyright © 2019, John Wiley and Sons). f Reproduced with permission (K. Brousse et al. 2020) (Copyright © 2019, John Wiley and Sons).

used to create templates of MSCs' electrode materials (Z. K. Wu et al. 2014). For example, as shown in Figure 8.2(a), Sun et al. (L. Sun et al. 2016) initially synthesized the SWNT/carbon current collectors with the help of the carbonization process, which converted the photoresist into SWNT coated by an amorphous substance.

Furthermore, the amalgamation of photolithography, along with reduction through pyrolysis, may also result in high performance MSCs, and such work has been reported by Hong et al. (X. Hong et al. 2017), Here the researchers synthesized quantum dots of tin/carbon (Sn/C QDs), acting as current collectors and also as the electrode material. Importantly, the Sn QDs possessing high conductivity allowed the manufactured MSC to offer a larger specific capacitance attaining a value of 5.79 mF cm^{-2}. The material could exhibit capacitance retention of 93.3% over 5,000 cycles, confirming its requisite cycling stability. Additional examples of executing more than only lithography to manufacture high-performance devices are reflected in the work of Tian et al. (X. Tian et al. 2016), as illustrated in Figure 8.2(b). They implemented photolithography for designing the patterns SiO_2 substrate of Ti/Au layer, along with techniques based on physical vapor deposition; this was followed by the growth of holey graphene with an intermediate layer of graphene oxide on the layer of current collectors implementing a drop-dry method, trailed by electro-polymerizing of the topmost polyaniline layer. The ingenious integration of these design approaches made it possible for the so-designed MSCs to deliver a large capacitance corresponding to the value of 271.1 $F.cm^{-3}$, large energy density corresponding to the value of 24.1 mWh cm^{-3} and appreciable cyclic stability. Overall, lithography is taken as an appropriate candidate for constructing several kinds of MSCs having precision and fair electrochemical properties. Attributed to the high resolution and robust operation process, lithography is mainly implemented in little integrated systems for causing their precision. However, attributed to the restrictions limited to this technology, it is required to be amalgamated with other approaches to effectively complete the fabrication process. In addition, the removal of the template is significant in buffer solution or at large temperatures, which might cause low efficiency and trouble for scaling up, restricting its development. The worst is that significant cost and the pollution caused by the photoresist reagents also lead to restricting its real application.

8.2.4 Laser Scribing

Scribing uses large-energy laser beams for illuminating the workpiece surface to melt the local area and gasify it, leading to scribing. As a result of its facile approach, large precision, and fast procedure, laser scribing is generally implemented in the fabrication of MSCs (J. Ye et al. 2018). For example, Peng et al. (Z. Peng et al. 2015) stated the design of porous graphene doped by boron in a facile way through laser induction. This was then transferred from the polyimide sheets and immersed in boric acid. Following the assembling through implementation of a solid electrolyte, the solid-state flexible MSC is available to be readily fabricated, and the manufactured devices exhibit a large areal capacitance corresponding to the value of 16.5 mF cm^{-2} which is three folds as compared to the devices fabricated with the help of electrode materials, which are non-doped. Similarly, Shi et al. (X. Shi et al. 2019) embarked on a one-step, scalable, and cost-effective production of LIG micropatterns using laser scribing (Figure 8.2(c)). Importantly, the process of laser scribing meant for commercial polyimide (PI) membrane is able to attain simultaneously the patterning and designing of LIG films (Figure 8.2(d) and (e)), which became very appropriate for large integrated MSCs in the absence of any metal interconnects, extra substrate, and current collectors. The LIG-MSCs implementing the gel electrolyte of PVA/H_3PO_4 could exhibit an areal capacitance of 0.62 mF cm^{-2} and long cyclic life in the absence of any degradation of capacitance over 10,000 cycles. Furthermore, this method also refrained the high cost and complexity associated with the traditional manufacturing process and stayed more desirable to the MSCs' practical applications. Furthermore, Li et al. (L. Li et al. 2016) designed flexible device possessing outstanding electrochemical performance by combining electrodeposition and using the laser-induction approach. That type of integration proposes synthesis of larger performance MSCs. Figure 8.2(f) shows yet another example, where Brousse et al. (K. Brousse et al. 2020) attempted the designing of flexible MSCs based on RuO_2 on Pt foil by implementing a modest approach by laser-writing of a bi-layered film. As a result of the pillar morphology of electrodes, the designed MSC shows a specific capacitance of 27 $mF.cm^{-2}$/540 $F.cm^{-3}$ in the electrolyte comprising of 1 M H_2SO_4, in combination with high cyclic life. In another instance, Jiang et al. (K. Jiang et al. 2020) attempted a facile laser-scribing approach applied on high-area polypyrrole-based film with benzene-bridge obtained through polymerization, which was interfacial in nature. The fabricated MSCs when tested in 1-ethyl-3-methylimidazolium tetrafluoroborate attained a power density of 9.6 kW cm¯3 and energy density of up to 50.7 mWh cm^{-3}. In brief, dissimilar to photolithography, laser scribing refrains from the application of the extra templates and complex processes. Furthermore, commercial polymer films can be transformed into graphene straightaway and further active materials, and this approach can also lead to the synthesis of active materials, such as metal oxides. Hence, it has seriously been considered in the designing of energy-storage devices of micro scale and has gathered a special consideration as an effective fabrication method to design high-capacitance MSCs. Furthermore, laser scribing imparts considerable tensile properties in the fabricated MSCs, in accordance with the work of Jiao et al. (Jiao et al. 2019). Therefore, laser scribing has gathered huge attention with respect to the fabrication of various kind of MSC devices.

8.2.5 Mask-Assisted Filtering

This is a method where micro energy devices are fabricated with the help of vacuum filtration of the electrode material in liquid form in combination with template assistance. This method is considered to be more brief, facile-operated, and cost-effective than the other electrode designing methods. Recently, Huang et al. (X. Huang and Wu 2020) described a very-effective and facile way for exfoliating MXene with several layers, implementing a mild water-freezing-and-thawing (FAT) approach, and enormous FAT-MXene flakes (reaching 39%) were acquired after four cycles of such approach. Following this process, the FAT-MXene could be applied to fabricate on-chip MSC based in all-MXene by the means of template-assisted filtering (Figure 8.3(a)). Due to the large amount of FAT-MXene related in this fabrication approach, the designed MSC showed a volumetric capacitance of 591 F cm^{-3} and areal capacitance of 23.6 mF cm^{-2} at the optimal thickness of electrode, as shown in Figure 8.3(b). In addition, it is simple and useful to fabricate asymmetrical MSCs (AMSCs) by implementing mask-assisted filtering. For instance, as shown in Figure 8.3(c), Qin et al. (J. Qin et al. 2019) initially used graphene current collectors accumulated on the mask, followed by adding the solution containing mesoporous MnO_2 nanosheets and the porous VN nanosheets solution onto different sides of the mask. Following a facile deposition method with mask-assisted approach was gel-electrolyte coating, leading to the procurement of all-solid state AMSC, together with porous VN nanosheets as the negative electrode and mesoporous MnO_2 nanosheets as positive electrode. The as-prepared AMSC exhibited a very high

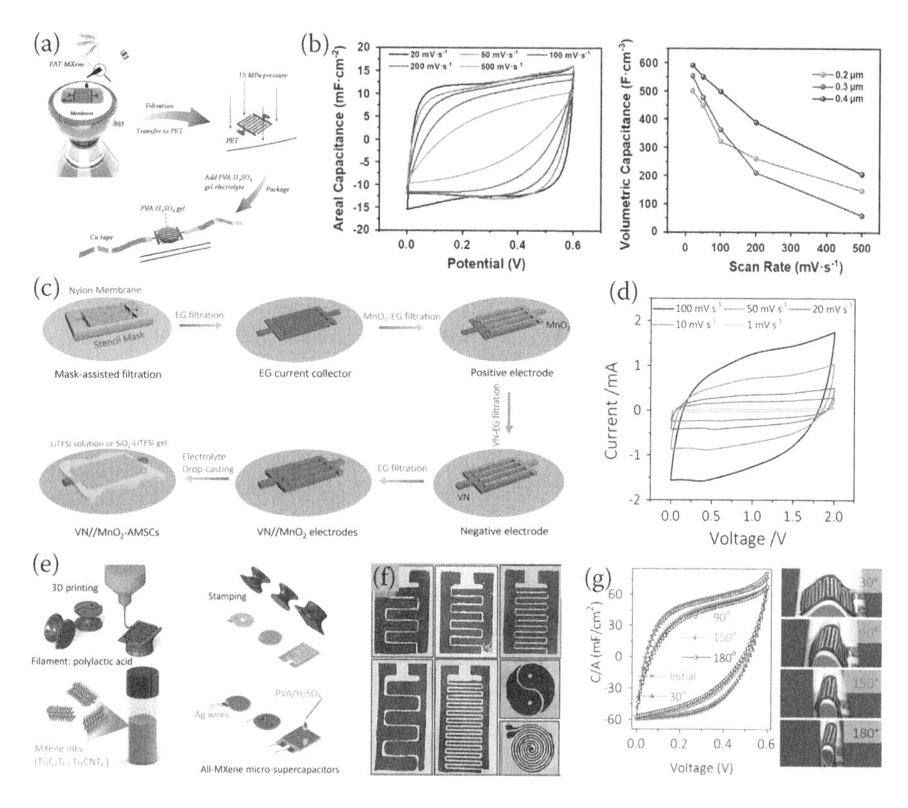

FIGURE 8.3 (a) The schematic corresponding to the mask-assisted fabrication of MSC. (b) Cyclic voltammetry plots at various scan rates (left) and the corresponding volumetric capacitances associated MSCs. (c) Fabrication by mask-assisted filtration. (d) Cyclic voltammetry plots obtained at various scan rates. (e) Fabrication methodology through the stamping strategy. (f) Digital photographs of the MSCs with different architectures. (g) Cyclic voltammetry plots of as-stamped MSCs associated various deformed states.

Source: (a, b) Reproduced with permission (J. Qin et al. 2019) (Copyright © 2020, John Wiley and Sons). (c, d)

volumetric energy density of 21.6 mWh.cm^{-3}, long-term cycle life in combination with a voltage window of 2 V (Figure 8.3(d)). Mask-assisted filtering is an effective way to fabricate MSCs due to its efficiency and convenience to grow both active materials and current collectors. Furthermore, it refrains from the requirement of expensive equipment or additional ink preparation, hence paving an easy pathway to retrieve efficiently performing MSCs. Nevertheless, the fabrication of masks is a very time-taking task, thereby restricting the development and application on a large scale. Furthermore, the substrate of MSCs retrieved by mask-assisted filtration is limited to filter membrane. Addition processes for transfer are generally necessary to transfer MSCs to other substrates. This process leads to having a direct impact on restraint of the selection of substrates and consumption of the preparation time.

8.3 MATERIALS DEPENDENT DESIGN PARAMETERS

8.3.1 Design of Wlectrode Materials

MSCs generally comprise four parts: electrolyte, electrodes, current collector, and substrate. Electrodes are basically brought into place to pave the path for adsorption sites for electrons and ions, either by the creation of electric double layers or through electrochemical reaction. Electrolyte crucially acts as element for the transportation of supplement ions to function ionic conduction intrinsic to the circuit. The accumulator (collector) is the portion responsible for linking the electrodes with the external wires, creating unblocked pathways for electrons passage to and fro among the external circuit and the electrodes. The substrate plays the role of a carrier for the entire MSC device. Among these parts, the electrode is considered as the most important component that can significantly affect the MSC performance (Xiong et al. 2014). Currently, the popularly implemented electrode materials can be classified into three categories: carbon materials, conducting polymers, and metal-based materials (S. Zheng et al. 2020). These not only possess diverse properties but also exhibit distinct charge-storage mechanisms (Najib and Erdem 2019). Presently, cogent selection and integration of various kinds of electrode materials that can be employed as functional to attain better electrochemical properties still remain at its infancy. Thus, this section discusses and compares the advantages and disadvantages of some important and typical electrode materials, along with discussing sensible resolutions to their drawbacks. Through this discussion, important standards for designing high-performance MSCs are outlined. Electric double-layer MSCs implementing carbon materials are broadly implemented electrodes for EDLCs credited to their high surface area, high compatibility, and large porosity with sophisticated manufacturing approaches of MSCs (Beidaghi and Wang 2012). Based on dimensionality, carbon materials are broadly divided into three types: zero-dimensional (0D) carbon materials such as graphene/carbon dots (W. W. Liu et al. 2013), active carbon (In et al. 2015), and one-dimensional (1D) carbon materials including CNTs and carbon nanofibers (Hsia et al. 2014) and two-dimensional (2D) carbon materials including graphene (A. K. Lu, Li, and Yu 2019). It is easy to estimate the fair rate capability and long cyclic life stability associated with the carbon electrodes from the point of view of the aforementioned merits. However, the kind of absorption mechanism that is electrostatic significantly limits their energy density. For this instance, different dimensional carbon materials possess their own separate remedies. For 1D and 0D carbon materials, the traditional strategy is the creation of porous nano/micro-structure to improve the active sites. For example, Li et al. (Yang Li et al. 2019) applied a solvent-free MOF-CVD method to form a layer of nanoporous carbon coat on the interdigitated gold electrodes, as shown in Figure 8.4(a).

Attributed to the high conductivity and large porosity of these MOF-derived porous carbon layer (Figure 8.4(b)), the fabricated MSCs brings a stack power of 233 $W.cm^{-3}$ at 1000 mV s^{-1}. It is estimated that the different kind of MOFs will also offer nanoporous carbon electrodes by similar methods. Apart from tuning of

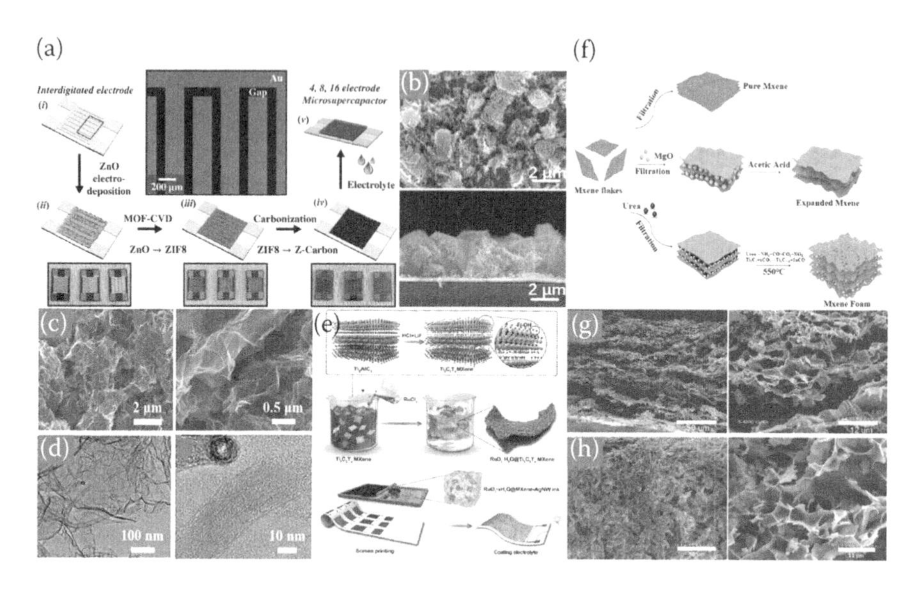

FIGURE 8.4 (a) Detailed illustration of designing approach associated with the MSCs. (b) Top-view and cross-sectional SEM images. (c) Top view and (d) cross-sectional SEM images. (e) Schematic illustration of the process for screen-printed MSCs. (f) Synthesis illustration of the designing of MXenes in various forms. Side views from SEM images of (g) expanded MXene and (h) foam of MXene.

Source: (a, b) Reproduced with permission (Copyright © 2019, Elsevier). (c, d) Reproduced with permission (Copyright © 2020, John Wiley and Sons). (e) Reproduced with permission (Copyright © 2019, John Wiley and Sons). (f–h) Reproduced with permission (Copyright © 2020, Elsevier).

porosity, deploying the composition of nano building blocks is another appreciable method of obtaining carbon electrodes with high-capacitance. For instance, Hsia et al. (Hsia et al. 2014) described the silicon wafer-based MSCs scaffolded on vertically aligned CNTs. The stability was enhanced by the perpendicularly arranged nanoarray, which result in significant suppression of the the agglomeration of the CNTs during electrode preparation. However, for 2D carbon materials, including graphene, numerous approaches can be undertaken to enhance its electrochemical properties. The most significant factor that limits its electrochemical properties is the 2D layer restacking during the preparation of the electrode. Integration of guest materials as spacers has projected itself to be a very effective way of preventing the stacking of 2D carbon materials. For example Wang et al. (Y. Wang et al. 2020) applied graphene with CNTs to design flexible MSCs. The restacking of the graphene sheets was prevented due to the presence of the CNTs (Figure 8.4(c) and (d)). Besides the aforementioned two strategies, atomic doping is another strategy that not only enhances the active sites but also improves the electrode materials' conductivity. Wu et al. (Z. S. Wu et al. 2014) designed a B and N co-doped graphene (BNG) film by a layer-by-layer assembly method and then tested it as a MSC, which delivered 488 F.cm^{-3}

of volumetric capacitance and good rate capability (2000 $mV.s^{-1}$). Such augmented electrochemical performance was attributed to the diatomic doping providing additional pseudo capacitance and also improving the wettability of graphene electrode.

8.3.2 Pseudo MSCs

Pseudocapacitive materials consist primarily of conducting polymers (Z. S. Wu et al. 2014) and metal-based (Ru, Co, Mn, and so on) compounds (Xiaofeng Wang et al. 2014). These materials store charge through reversible and rapid faradic reactions, causing a larger capacity than the carbon-based materials. In particular, conducting polymers have garnered huge potential for constructing MSCs due to their conductivity, eco-friendly status, and low cost. Despite having such credits, due to the undesirable agglomeration and instability of the polymer backbone, MSCs made of conducting polymer agonize from cyclic life and low energy density. It is efficient and sensible to distribute the infusion and electrolyte diffusion through rational structure design. For instance, Zhu et al. (M. Zhu et al. 2017) applied an electrochemical approach to promote the growth on a conductive glass of sac-like polypyrrole nanowires. In such a system, sac-like structures could facilitate the dissemination of electrolyte, which in turn caused improvement in the electrochemical performance of the material. Therefore, the so fabricated MSCs with polypyrrole nanowires exhibited 15.25 $mWh.cm^{-3}$ of energy density at the power density of 0.89 $W.cm^{-3}$. Even if metal-based materials are broadly used as electrode materials, their poor conductivity and smaller cyclic stability restricts their use. An effective approach is to dope them by atoms for improving the electrical conductivity, in turn leading to improvement in their electrochemical performance. As shown in Figure 8.4(e), Li et al. (Hongpeng Li et al. 2019) implemented RuO_2 as spacers to enhance the gap between the MXene sheets, and then combined it with silver nanowire (AgNW) to design electrodes for MSCs. The spacing imparted by the RuO_2 nanoparticles improved the ion channels, and the so-prepared MSC exhibited large volumetric capacitances of 864.2 F cm^{-3} at 1 mV s^{-1} and 304.0 F cm^{-3} at 2000 mV s^{-1}. The long-term cycling stability corresponded to 90% of initial capacitance retention over 10,000 cycles in combination with outstanding flexibility. Similarly, Chen et al. (X. Chen et al. 2019) integrated MoS_2 into MXene and tested the resulting material as electrode material for symmetrical MSCs. The implementation of MoS_2 not only limited the agglomeration of the MXene sheets but also resulted in additional pseudocapacitance. As a result, the fabricated MSC showed specific capacitance of 173.6 F cm^{-3} and energy density of 15.5 mWh cm^{-3}. Furthermore, as shown in Figure 8.4(f), Zhu et al. (Y. Zhu et al. 2020) attempted reduction of MXene layers from stacking by expanding the interlayer gap with foreign materials and generating in-plane pores (Figure 8.4(g) and (h)). Attributed to these matters, the ion transport efficiency and the electrochemically active area got largely improved, leading to much enhanced capacitive performance (Bu et al. 2020).

8.4 CURRENT STATUS

The perpetual development of the microelectronic is attributed to the progress of energy-storage devices (W. Li, Meredov, and Shamim 2019). MSC can suggestively satisfy the increasing requirements of flexible and largely integrated electronics credited to its extremely high charge/discharge rate, small-size, and lightweight, flexible nature. Till date, these have already been efficiently applied to progressive wearable devices, different kinds of sensors, and so on (Chen Zhao et al. 2018). Furthermore, these can significantly enhance the medicine-related area credited to the diversity of choice of electrode materials. The continued development in the arena of wearable devices and numerous little sensors have largely elevated the need for corresponding power supply modules (Pu, Hu, and Wang 2018). Nevertheless, for the moment, the significant trouble relies in receiving long-term continuous and stable power supply modules. The critical working conditions of solar cells (only with illumination) and small service life of micro-batteries are making it impossible for these devices to meet this requirement. Linking MSCs with energy harvest strategies (e.g., triboelectric nanogenerator) proves to be an attractive solution for meeting the requirement. For instance, Zhang et al. (S. L. Zhang et al. 2019) have illustrated the strategy by implementing double-energy harvest devices, e.g., triboelectric nanogenerator (TENG) and electromagnetic generators (EMG) to collect the mechanical energy (Figure 8.5(a)), in turn led to the increase of the charging rate of MSCs.

As shown in Figure 8.5(b), Jiang et al. (Q. Jiang et al. 2018) designed a compacted self-powered electric module from the MXene and TENG-MSC for wearing safely on human skin. The device transformed the mechanical energy as a result of the motion, into electrical energy and supply it to the MSC for storage. It is highly desirable to design multifunctional materials for the various functional devices. For instance, Qin et al. (J. Qin et al. 2020) attempted the synthesis of hierarchically ordered 2D dual-mesoporous Ppy/graphene (DM-PG) nanosheets as active materials for both NH_3 sensor and MSC (Figure 8.5(c)). Apart from gas sensor, Song et al. (Y. Song et al. 2018) also attempted the assemblage of the MSC and piezoelectric sensors (PRSs) for fabricating a real-time pressure sensor. It is noteworthy that both the sensor and the MSC worked well after integration. The PRS exhibited large sensitivity of 0.51 kPa^{-1} and broad detection range, while MSC showed excellent areal capacitance. In addition, it has been proposed to be implemented as a 3D touch. Guo et al. (R. Guo et al. 2017) reported a strategy to integrate nanowire-based UV sensor and MSCs on a piece of paper with printing Ni circuit. The resulting MSC-UV sensor system exhibited self-powering and good sensing capabilities. And the MSC could be charged by connecting it with a solar cell, as shown in Figure 8.5(e). Similarly, Cai et al. (Cai, Lv, and Watanabe 2016) showed a self-powered system of UV-light detector by combining a MSC designed by laser direct writing on PI films, a photodetector based on ZnO nano particles also fabricated by one-step direct laser writing and a solar panel.

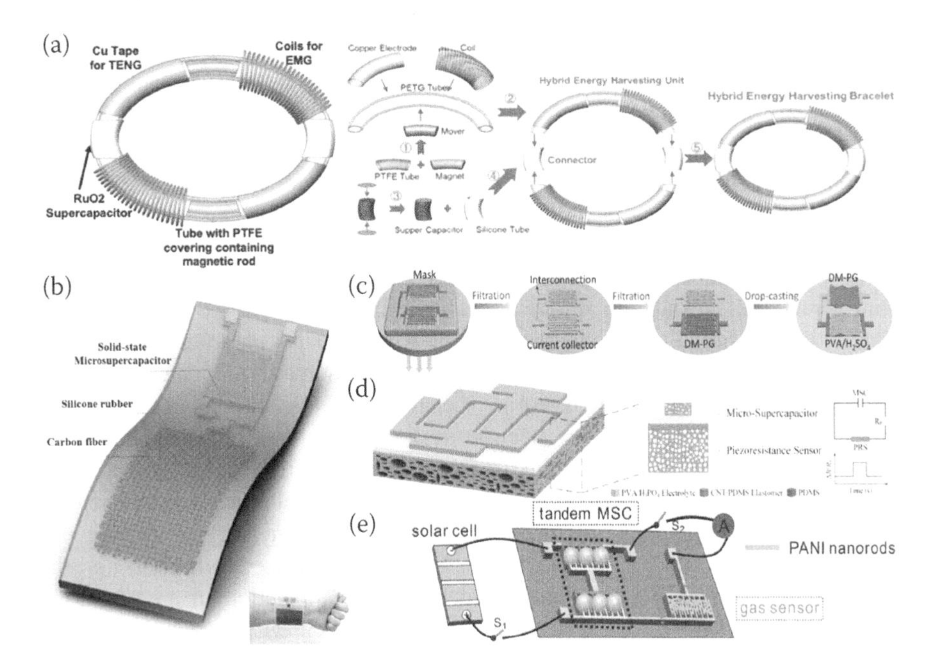

FIGURE 8.5 (a) The fabrication technique of the bracelet meant for energy harvesting. (b) Wearable self-charging system device, system worn on the forearm (inset). (c) Fabrication of the planar MSC-sensor system. (d) Schematic diagram of all-in-one sensing patch. (e) Schematic of the self-powered device.

Source: (a) Reproduced with permission (S. L. Zhang et al. 2019) (Copyright © 2019, John Wiley and Sons). (b) Reproduced with permission (Q. Jiang et al. 2018) (Copyright © 2018, Elsevier). (c) Reproduced with permission (J. Qin et al. 2020) (Copyright © 2020, John Wiley and Sons). (d) Reproduced with permission[132] (Copyright © 2018, Elsevier). (e) Reproduced with permission (R. Guo et al. 2017) (Copyright © 2017, John Wiley and Sons).

8.5 QUANTUM BATTERIES AND SUPERCAPACITORS

Quantum batteries (QB) and supercapacitors (QS) are defined as a large ensemble of n-dimensional quantum system with non-degenerate energy levels or non-passive state to reversibly extract or deposit the energy by unitary cyclic operations (Ferraro et al. 2019). Classically invented charge-storage devices are limited by their slow-speed operation, mainly attributed to the slow movement of heavy ions in a crystal lattice as well as in an electrolyte. Alternatively, the charge transfer occurs due to mechanical moving of ions or the interface phenomena at the surface of electrodes. Moreover, the operation of supercapacitors is limited to only 2–3 V, and a large increase in the leakage current further reduces an energy-storage time (Mishra et al. 2019). Therefore, QB/QS systems are essential to have compatibility with the ultrahigh speed integrated circuits as elements of memory with the implementation of the new quantum strategies (Ferraro et al. 2018). Quantum mechanics employs tools such as exchange and correlations in n-dimensional electron system with long-range interactions for engineering of energy-storage applications (Santos et al. 2019).

So far, there are a few models presented for the experimental demonstrations, such as spin-chains, nanofabricated quantum dots coupled to cavities, and superconducting qubits (Kai Xu et al. 2020; Baart et al. 2016) (Heinrich et al. 2021). There are also reports on the formation of two-dimensional electron double layers in van der Waals heterostructures comprised of graphene-MoS_2, WSe_2 and hetero-interfaces of metal oxides (Naik and Rabinal 2020; G. H. Lee et al. 2014).

One of the main challenges of current technology is represented by energy storage. In this scenario, many devices like capacitors and batteries are involved, from personal electronics to transport sector. Supercapacitors are improved versions of conventional capacitors that exploit a molecular-scale interface between the ions of an electrolyte and the electrode to increase the energy density while displaying large power densities. Quantum effects, such as exchange and correlations in low-dimensional electron systems with long-range Coulomb interactions, constitute powerful tools that can be potentially manipulated and engineered for energy-storage applications. For general mesoscopic device, the electronic contribution C_e to the capacitance of a mesoscopic device can be written as $C_e^{-1} = C_g^{-1} + C_q^{-1}$, where C_g is a classical contribution, i.e., the conventional geometric capacitance, and C_q is a contribution, from "quantum capacitance" (Bueno 2019). The latter accounts for the variation of the Fermi energy due to charge accumulation. Usually, quantum contribution has the net effect to lower the capacitance of the device, thereby reducing the stored energy density with respect to the classical case. However, situations may arise where a negative exchange and correlation contribution to the energy dominates over the positive kinetic energy do exist (Büttiker, Thomas, and Prêtre 1993), leading to $C_q<0$ and $C>C_g$. Such quantum mechanical enhancement of the total capacitance as compared to the classical value has been observed in several systems, including two-dimensional electron double layers formed in GaAs semiconductor quantum wells (Büttiker, Thomas, and Prêtre 1993), the interface between two oxides ($LaAlO_3/SrTiO_3$) (Tao et al. 2020), 2D monolayers of WSe_2 (Bera et al. 2019), and graphene-MoS_2 heterostructure (Lili Yu et al. 2014).

Giuliani and Vignale (Giuliani and Vignale 2005) in their book interpreted the compressibility of the two-dimensional electron liquid as a capacitance by calculating charge ρ per unit area of the gate, i.e., the capacitance per unit area, as shown in Figure 8.6.

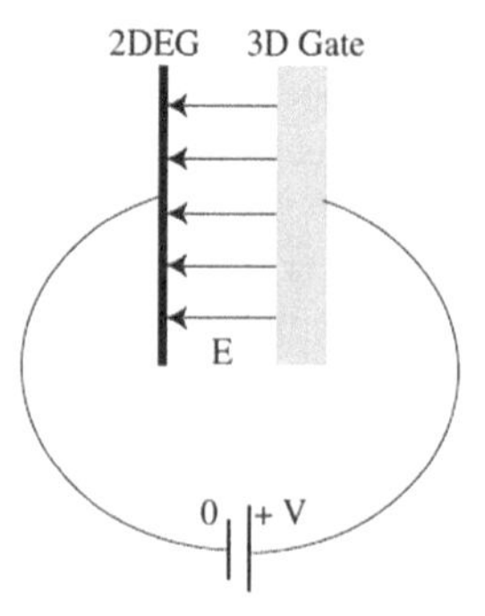

FIGURE 8.6 Capacitance measurement between a three dimensional gate and a two dimensional electron liquid.

Source: Reprinted with permission from (Giuliani and Vignale 2005), Copyright © 2005, Cambridge University Press.

The differential capacitance C/A = $\delta\rho/\delta V$ is given by

$$\frac{C}{A} = \frac{\epsilon_b}{4\pi(d + \lambda)} \tag{8.1}$$

where, $\delta V = \delta Ed + \frac{\partial \mu}{\partial n}\frac{\partial \rho}{e^2}$, $\delta E = \frac{4\pi\delta\rho}{\epsilon_b}$, and screening length $\lambda = \frac{\epsilon_b}{4\pi e^2}\frac{\partial \mu}{\partial n}$.

It is clear that the capacitance value greatly depends on the screening length. Although the separation between the plates is in microscopic scale, the screening length is in the order of Bohr radius. Thus, the capacitance determination has to be done through a sensitive measurement only. Any change in the chemical potential will lead to the change in the field. Recently, quantum-mechanical effects are exploited to enhance the energy storage in supercapacitors (Ferraro et al. 2019). Ferraro et al. published a model for the quantum supercapacitor by considering two chains of double quantum dots, one containing electron and other holes (Figure 8.7) (Ferraro et al. 2019). This basic model is used as designing a building block for realizing charge and spin qubits in experimental architectures. Double quantum dots is a qubit.

Screened Coulomb interactions between electron and hole in an embedded photonic cavity was shown responsible for long-range coupling between all the qubits. Therefore, this study demonstrated an improvement in the energy transfer and storage performance through purely quantum mechanical effects.

A patent on quantum supercapacitor by Alexandr Mikhailovich Ilyanok, in 2006 (Ilyanok 2006), used nanostructured materials between the two plates (Figure 8.8). The sizes of the clusters vary from 7–29 nm. The energy in the supercapacitor is stored uniformly in the material, thus inducing a resonant coupling of the electrons. In this system, the maximum stored specific energy was 1.66 MJ/kg. Thus, the architecture of the cells can be used for both integrated circuits and energy storages in combination. In this design, the extreme achievable current density in quantum supercapacitor is:

$$j_e = \frac{ef_e}{\pi r_0^2} = \frac{4\pi e m_e^3 \alpha^8 c^4}{h^3} = 3.4 \cdot 10^4 A/cm^2 \tag{8.2}$$

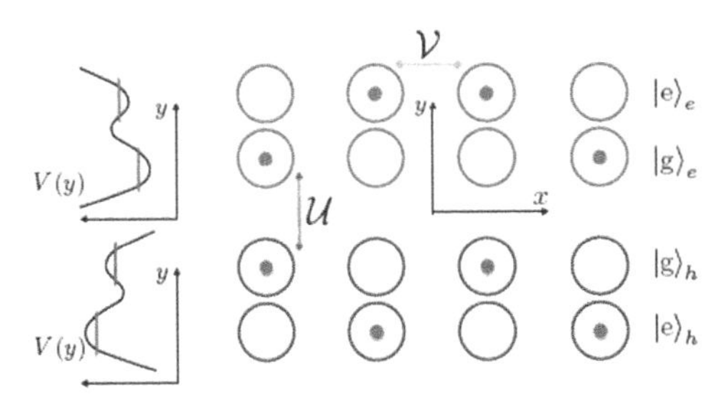

FIGURE 8.7 Schematic of the two-chain system proposed by Ferroro et al.

Source: Reprinted with permission from (Ferraro et al. 2019), Copyright © 2019, American Physical Society.

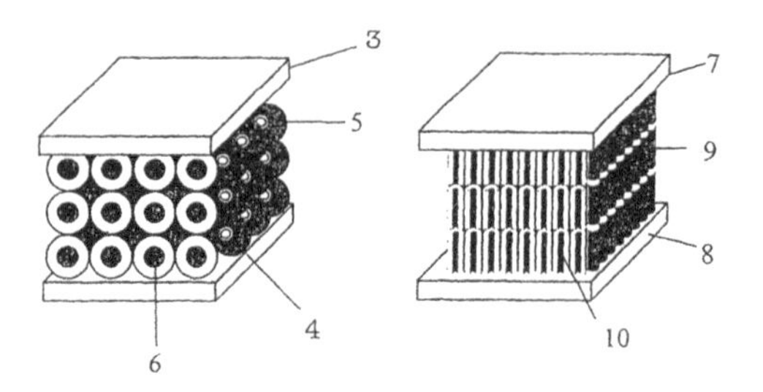

FIGURE 8.8 The quantum supercapacitor design dielectric comprising central-symmetric clusters.

The capacitance of supercapacitors mainly depends on extraordinarily high electrical conductivity and high surface area, and hence, graphene-based materials outstand with great potential for supercapacitor applications. These act as electrical double-layer capacitors with a high power density and excellent cycling capability (Garakani et al. 2021). Extensive studies have been done in graphene for this application. Graphene is used in several architectures, including zero-dimensional quantum dots, one-dimensional fibers, two-dimensional films, and three-dimensional porous structures employed for supercapacitor electrodes. Graphene quantum dots (GQDs) present unique properties owing to quantum confinement and edge effects (De et al. 2020). GQDs are expected to provide enhanced electrochemical double-layer capacitance and emit stronger luminescence. Another variant of GQDs is graphene quantum dot-decorated porous carbon spheres that have shown an extensive potential for an efficient supercapacitor (Y. Deng et al. 2020). The GQDs modify the conductivity of the resultant materials.

Alongside graphene, another technology for making supercapacitors makes asymmetric electrode supercapacitors. Asymmetric hybrid devices are developed in which the cathode electrode is a pseudo-capacitive oxide electrode and the activated carbon electrode as the anode for an electrical double-layer to provide larger voltage window and higher specific energy. A facile microwave-assisted hydrothermal synthesis of SnO_2 quantum dots was demonstrated as an active material for solid-state asymmetric supercapacitor (Geng et al. 2020). $NiCo_2S_4$ quantum dots have shown also similar performance as asymmetric devices (Wenyong Chen et al. 2020). P-N heterojunction $LaMoO_3$/GQD can act as negative electrode for asymmetric supercapacitor (Z. Shi et al. 2020). Some other example has been observed also. Nb_2O_5 quantum dots coated with biomass carbon can also act for ultra-stable lithium-ion supercapacitors (Lian et al. 2021). Graphene/reduced graphene oxide as electrodes have also shown super capacitive behavior.

9 Future Applications

Supercapacitors find use in consumer electronic devices, hybrid electric vehicles, buses, trains, and industrial applications. These are used in various ways; for smaller devices, they are used as a charge storage, and larger equipment use them to facilitate a quick start. In the future, supercapacitors will compliment batteries in many applications for their low temperature operation performance, cycle life, fast charge/discharge, and possibility to be used in compact forms. Supercapacitors have an excellent potential for portable applications, such as power electronics systems, due to their high capacitance values up to 2700 F, 10 times lower equivalent series resistance than conventional capacitors and long cycle lives. The next generation applications of supercapacitors based on current advancements are summarized in the following sections.

9.1 POWERING SMALL ROBOTS

The supercapacitors can be used in future robots fully or in hybrid mode with batteries. The characteristic advantages of supercapacitors over batteries, such as high power density, high cyclic life, and storage of regenerative power from a moving robot during braking or downhill motion, can be utilized efficiently. The solar-powered supercapacitors can be used in powering the small robots in outside applications. Artal et al. developed an autonomous mobile robot named 'MBot,' which uses a supercapacitor as its power source (Artal, Bandrés, and Fernández 2011). Additionally, the growth of carbon-based electrode materials such as graphene, CNT, and activated carbon, encourages the scientist to use high energy density supercapacitors in robotic applications.

9.2 FLEXIBLE AND WEARABLE ELECTRONICS

The development of hydrogels with high electrical conductivity, transparency, stretchability, and mechanical strength makes them useful as electrode material for supercapacitors. The aesthetics of these supercapacitors allow them to be used in powering the wearable sensors, implants, and electronic systems for humans and robots. The hydrogels soaked with electrolytes facilitate electrical double-layer formation at metal-hydrogel and hydrogel-elastomer interface. The voltage application triggers the motion of electrons/ions toward or away from the interface causing polarization at the hydrogel interface. This results in squeezing of the elastomer actuator in artificial muscle (C. Yang and Suo 2018). Figure 9.1 shows a fabric-based flexible supercapacitor device with high stability and capacitance retention under bending and wet conditions. The supercapacitors are utilized in self-powered pressure sensors in artificial skins. Here, the hydrogel layer remains

DOI: 10.1201/9781003174554-9

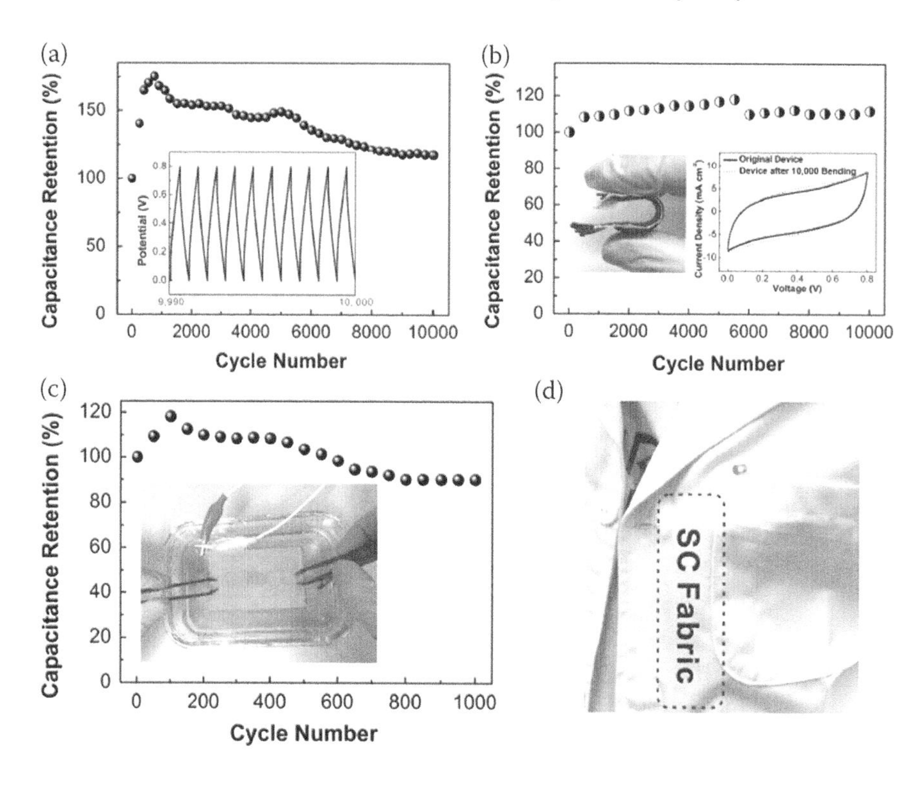

FIGURE 9.1 Fabric-based wearable supercapacitor device (a) stability test-based on GCD curve at 20 mA/cm^2, (b) capacitance stability under bending, (c) capacitance retention under water, and (d) digital image showing supercapacitor fabric device sown over lab coat.

Source: Reprinted with permission from (Yu Yang, Huang, et al. 2017). Copyright © 2017, John Wiley and Sons.

connected to the capacitive meter, and on application of pressure, the total capacitance of the system changes. The development of conductive flexible substrates has been the focus of researchers.

The conductive and stretchable substrates allow the utilization of supercapacitors to power the wearable sensors and devices. He et al. have developed CNT-based high performance stretchable supercapacitor without using any elastic material (S. He et al. 2016). They have modified the design of the supercapacitor by using interconnected segments. The advantage of flexible supercapacitors is the ability to withstand high strain, shape change, and light weight.

9.3 SUPERCAPACITORS IN RENEWABLE ENERGY DEVICES

The future of mankind depends upon the reusability of the available clean and renewable energy resources, such as solar, wind, and tidal energy. The systems for energy harvesting from above-mentioned resources require a sustainable energy-storage system due to lack of continuous supply. The supercapacitors have

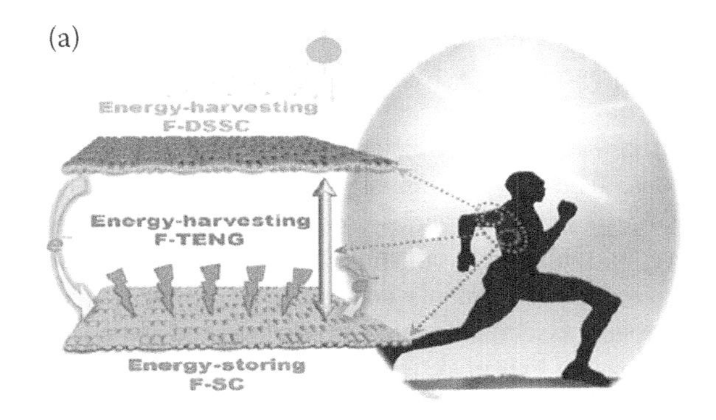

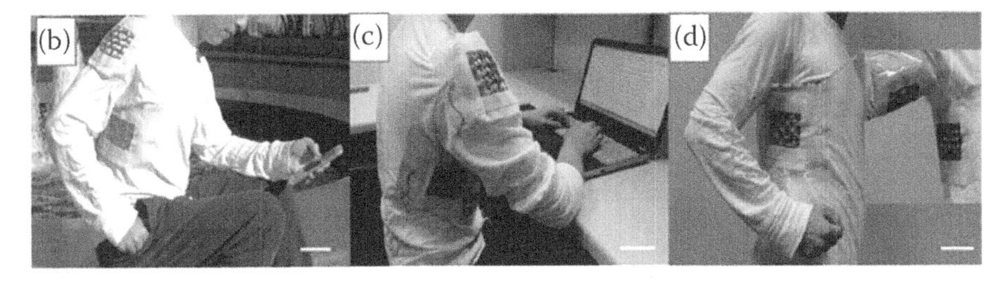

FIGURE 9.2 (a) Schematic representation of solar energy utilization to charge the supercapacitor device that is integrated with fabric-based dye-sensitized solar cell (F-DSSC). (b), (c), and (d) demonstration of solar-powered device at outdoor, indoor, and working conditions.

Source: Reprinted with permission from (K. and Rout 2021), Copyright © 2021, Royal Society of Chemistry.

advantage over batteries due to high cyclic life and high power density. The supercapacitors become an ideal choice to be integrated with above-mentioned renewable energy resources considering the fluctuation in voltage and frequent charging/discharging requirements (Abbey and Joos 2007). The electric vehicles (EVs) are now utilizing the hybrid mechanism of energy supply from battery and supercapacitors. Solar-powered supercapacitors are the main focus of EV manufacturers due to ease in installation of rooftop solar panels that can charge the supercapacitors (K. and Rout 2021). Figure 9.2 shows the schematic of utilization of solar energy to charge a photo-powered supercapacitor via a dye-sensitized solar cell (K. and Rout 2021). The utilization of supercapacitors in hybrid systems reduces weight and the overall size of the power-storage systems. The utilization of solar light enables wireless charging of the device, which is getting huge appreciation in wireless medical sensors.

9.4 CHALLENGES OF SUPERCAPACITOR UTILIZATION IN FUTURE APPLICATIONS

There are several challenges associated with the supercapacitor applications that scientists need to address. The major issue is with the lower energy density of the supercapacitors in comparison with batteries. This restricts their utilization as single-energy storage devices for a system. The lower operating voltage of supercapacitors requires multiple devices to be arranged in series to meet the voltage requirement. The high self-discharge rate also creates problems in actual applications. The environmental concerns related to ionic or organic electrolytes in large numbers of series-connected devices poses a real concern. Bio-compatibility of the device constituents is also a big challenge in using the supercapacitors for bio-implants and wearable applications. The reliability and repeatability issues associated with self-powered sensors needs to be assessed properly. However, using hybrid systems with battery-like electrodes and pseudo-capacitive electrode materials is solving most of the energy-density issues, but their compatibility with future applications needs to be studied if full utilization is desired.

References

Abbey, Chad, and Géza Joos. 2007. "Supercapacitor Energy Storage for Wind Energy Applications." *IEEE Transactions on Industry Applications* 43 (3): 769–76. doi:10.1109/TIA.2007.895768

Abdolahi, Ahmad, Esah Hamzah, Zaharah Ibrahim, and Shahrir Hashim. 2012. "Synthesis of Uniform Polyaniline Nanofibers through Interfacial Polymerization." *Materials* 5 (8): 1487–94. doi:10.3390/ma5081487

Abdolhosseinzadeh, Sina, René Schneider, Anand Verma, Jakob Heier, Frank Nüesch, and Chuanfang Zhang. 2020. "Turning Trash into Treasure: Additive Free MXene Sediment Inks for Screen-Printed Micro-Supercapacitors." *Advanced Materials* 32 (17): 1–9. doi:10.1002/adma.202000716

Abouali, Sara, Mohammad Akbari Garakani, Biao Zhang, Zheng Long Xu, Elham Kamali Heidari, Jian Qiu Huang, Jiaqiang Huang, and Jang Kyo Kim. 2015. "Electrospun Carbon Nanofibers with in Situ Encapsulated Co3O4 Nanoparticles as Electrodes for High-Performance Supercapacitors." *ACS Applied Materials and Interfaces* 7 (24): 13503–11. doi:10.1021/acsami.5b02787

Alabadi, Akram, Shumaila Razzaque, Zehua Dong, Weixing Wang, and Bien Tan. 2016. "Graphene Oxide-Polythiophene Derivative Hybrid Nanosheet for Enhancing Performance of Supercapacitor." *Journal of Power Sources* 306: 241–47. doi:10.1016/j.jpowsour.2015.12.028

Ali, Mai, Lutfi Albasha, and Hasan Al-Nashash. 2011. "A Bluetooth Low Energy Implantable Glucose Monitoring System." *European Microwave Week 2011: "Wave to the Future", EuMW 2011, Conference Proceedings – 41st European Microwave Conference, EuMC 2011*, no. October. IEEE: 1265–68.

Ambade, Rohan B., Swapnil B. Ambade, Rahul R. Salunkhe, Victor Malgras, Sung Ho Jin, Yusuke Yamauchi, and Soo Hyoung Lee. 2016. "Flexible-Wire Shaped All-Solid-State Supercapacitors Based on Facile Electropolymerization of Polythiophene with Ultra-High Energy Density." *Journal of Materials Chemistry A* 4 (19): 7406–15. doi:10.1039/c6ta00683c

Antiohos, Dennis, Glenn Folkes, Peter Sherrell, Syed Ashraf, Gordon G. Wallace, Phil Aitchison, Andrew T. Harris, Jun Chen, and Andrew I. Minett. 2011. "Compositional Effects of PEDOT-PSS/Single Walled Carbon Nanotube Films on Supercapacitor Device Performance." *Journal of Materials Chemistry* 21 (40): 15987–994. doi:10.1039/c1jm12986d

Ardizzone, S., G. Fregonara, and S. Trasatti. 1990. "'Inner' and 'Outer' Active Surface of RuO2 Electrodes." *Electrochemica Acta* 35: 263–67. doi:10.1109/icdm.2002.1184004

Aricò, Antonino Salvatore, Peter Bruce, Bruno Scrosati, Jean-Marie Tarascon, and Walter Van Schalkwijk. 2010. "Nanostructured Materials for Advanced Energy Conversion and Storage Devices." *Materials for Sustainable Energy* 4: 148–59.

Artal, J.S., R. Bandrés, and G. Fernández. 2011. "Ulises: Autonomous Mobile Robot Using Ultracapacitors-Storage Energy System." *International Conference on Renewable Energies and Power Quality* 1: 1105–10.

Augustyn, Veronica, Patrice Simon, and Bruce Dunn. 2014. "Pseudocapacitive Oxide Materials for High-Rate Electrochemical Energy Storage." *Energy and Environmental Science* 7 (5): 1597–614. doi:10.1039/c3ee44164d

Baart, Timothy Alexander, Takafumi Fujita, Christian Reichl, Werner Wegscheider, and Lieven Mark Koenraad Vandersypen. 2016. "Coherent Spin-Exchange via a Quantum Mediator." *Nature Nanotechnology* 12 (1): 26–30. doi:10.1038/nnano.2016.188

Bagheri, Narjes, Alireza Aghaei, Mohammad Yeganeh Ghotbi, Ehsan Marzbanrad, Nick Vlachopoulos, Leif Häggman, Michael Wang, et al. 2014. "Combination of Asymmetric Supercapacitor Utilizing Activated Carbon and Nickel Oxide with Cobalt Polypyridyl-Based Dye-Sensitized Solar Cell." *Electrochimica Acta* 143: 390–97. doi:10.1016/j.electacta.2014.07.125

Banerjee, Joyita, Kingshuk Dutta, M. Abdul Kader, and Sanjay K. Nayak. 2019. "An Overview on the Recent Developments in Polyaniline-Based Supercapacitors." *Polymers for Advanced Technologies* 30 (8): 1902–21. doi:10.1002/pat.4624

Banerjee, Soma, and Kamal K. Kar. 2020. "Conducting Polymers as Electrode Materials for Supercapacitors." *Springer Series in Materials Science* 302: 333–52. doi:10.1007/978-3-030-52359-6_13

Banik, R., N. Suresh, N.R. Mandre, and Mineral Engineering. 2010. "Mineral Processing and Extractive Metallurgy Review: An International Journal." no. May 2013: 37–41.

Bard, Allen J., and Larry R. Faulkner. 2001. *Electrochemical Methods: Fundamentals and Applications*. 2nd ed. John Wiley & Sons, Inc.

Becker, Howard I. 1957. Low voltage electrolytic capacitor, issued July 23, 1957. https://patents.google.com/patent/US2800616A/en

Béguin, François, Volker Presser, Andrea Balducci, and Elzbieta Frackowiak. 2014. "Carbons and Electrolytes for Advanced Supercapacitors." *Advanced Materials* 26 (14): 2219–51. doi:10.1002/adma.201304137

Beidaghi, Majid, and Yury Gogotsi. 2014. "Capacitive Energy Storage in Micro-Scale Devices: Recent Advances in Design and Fabrication of Micro-Supercapacitors." *Energy and Environmental Science* 7 (3): 867–84. doi:10.1039/c3ee43526a

Beidaghi, Majid, and Chunlei Wang. 2012. "Micro-Supercapacitors Based on Interdigital Electrodes of Reduced Graphene Oxide and Carbon Nanotube Composites with Ultrahigh Power Handling Performance." *Advanced Functional Materials* 22 (21): 4501–10. doi:10.1002/adfm.201201292

Bera, M.K., R. Kharb, N. Sharma, A.K. Sharma, R. Sehrawat, S.P. Pandey, R. Mittal, and D.K. Tyagi.2019."Influence of Quantum Capacitance on Charge Carrier Density Estimation in a Nanoscale Field-Effect Transistor with a Channel Based on a Monolayer WSe 2 Two-Dimensional Crystal Semiconductor." *Journal of Electronic Materials*. Springer New York LLC. doi:10.1007/S11664-019-07058-0, 48: 3504–3513.

Bharathi Sankar Ammaiyappan, A., and Seyezhai Ramalingam. 2021. "Self-Powered Supercapacitor for Low Power Wearable Device Applications." *IOP Conference Series: Earth and Environmental Science* 850 (1): 012016. doi:10.1088/1755-1315/850/1/012016

Bo, Zheng, Changwen Li, Huachao Yang, Kostya Ostrikov, Jianhua Yan, and Kefa Cen. 2018. "Design of Supercapacitor Electrodes Using Molecular Dynamics Simulations." *Nano-Micro Letters* 10 (2): 1–23. doi:10.1007/s40820-018-0188-2

Boos, D.L. 1968. Electrolytic capacitor having carbon paste electrodes, issued May 29, 1968. patent no. 3536963.

Brezesinski, Torsten, John Wang, Sarah H. Tolbert, and Bruce Dunn. 2010. "Ordered Mesoporous α-MoO3 with Iso-Oriented Nanocrystalline Walls for Thin-Film Pseudocapacitors." *Nature Materials* 9 (2): 146–51. doi:10.1038/nmat2612

Brousse, K., S. Nguyen, A. Gillet, S. Pinaud, R. Tan, A. Meffre, K. Soulantica, et al. 2018. "Laser-Scribed Ru Organometallic Complex for the Preparation of RuO2 Micro-Supercapacitor Electrodes on Flexible Substrate." *Electrochimica Acta* 281 (August): 816–21. doi:10.1016/J.ELECTACTA.2018.05.198

Brousse, Kévin, Sébastien Pinaud, Son Nguyen, Pier Francesco Fazzini, Raghda Makarem, Claudie Josse, Yohann Thimont, et al. 2020. "Facile and Scalable Preparation of Ruthenium Oxide-Based Flexible Micro-Supercapacitors." *Advanced Energy Materials* 10 (6): 1–9. doi:10.1002/aenm.201903136

Brousse, Thierry, Daniel Bélanger, and Jeffrey W. Long. 2015. "To Be or Not To Be Pseudocapacitive?" *Journal of the Electrochemical Society* 162 (5): A5185–189. doi:10.1149/2.0201505jes

Bu, Fan, Weiwei Zhou, Yihan Xu, Yu Du, Cao Guan, and Wei Huang. 2020. "Recent Developments of Advanced Micro-Supercapacitors: Design, Fabrication and Applications." *Npj Flexible Electronics* 4 (1): 1–16. doi:10.1038/s41528-020-00093-6

Bueno, Paulo R. 2019. "Nanoscale Origins of Super-Capacitance Phenomena." *Journal of Power Sources* 414 (February): 420–34. doi:10.1016/J.JPOWSOUR.2019.01.010

Burke, Laurence D., and Michael E.G. Lyons. 1986. "Electrochemistry of Hydrous Oxide Films." *Modern Aspects of Electrochemistry*, no. 18: 169–248. doi:10.1007/978-1-4613-1791-3_4

Büttiker, M., H. Thomas, and A. Prêtre. 1993. "Mesoscopic Capacitors." *Physics Letters A* 180 (4–5): 364–69. doi:10.1016/0375-9601(93)91193-9

Cai, Jinguang, Chao Lv, and Akira Watanabe. 2016. "Laser Direct Writing of High-Performance Flexible All-Solid-State Carbon Micro-Supercapacitors for an on-Chip Self-Powered Photodetection System." *Nano Energy* 30: 790–800. doi:10.1016/j.nanoen.2016.09.017

Cao, Peng, Yuxia Fan, Junrui Yu, Rongmin Wang, Pengfei Song, and Yubing Xiong. 2018. "Polypyrrole Nanocomposites Doped with Functional Ionic Liquids for High Performance Supercapacitors." *New Journal of Chemistry* 42 (5): 3909–16. doi:10.1039/c7nj04367h

Cazade, Pierre Andre, Remco Hartkamp, and Benoit Coasne. 2014. "Structure and Dynamics of an Electrolyte Confined in Charged Nanopores." *Journal of Physical Chemistry C* 118 (10): 5061–72. doi:10.1021/jp4098638

Channu, Venkata Subba Reddy, Rudolf Holze, Scott Ambrose Wicker Sr., Edwin H. Walker Jr., Quinton L. Williams, and Rajamohan R. Kalluru. 2011. "Synthesis and Characterization of (Ru-Sn)O₂ Nanoparticles for Supercapacitors." *Materials Sciences and Applications* 02 (09): 1175–79. doi:10.4236/msa.2011.29158

Chapman, David Leonard. 2010. "LI. A Contribution to the Theory of Electrocapillarity." *The London, Edinburgh, and Dublin Philosophical Magazine and Journal of Science*, Philosophical Magazine Series 6, 25:148, 475–481. April. 1913, doi:http://dx.doi.org/10.1080/14786440408634187

Chen, Chuan Rui, Haili Qin, Huai Ping Cong, and Shu Hong Yu. 2019. "A Highly Stretchable and Real-Time Healable Supercapacitor." *Advanced Materials* 31 (19): 1–10. doi:10.1002/adma.201900573

Chen, Hsin Wei, Chih Yu Hsu, Jian Ging Chen, Kun Mu Lee, Chun Chieh Wang, Kuan Chieh Huang, and Kuo Chuan Ho. 2010. "Plastic Dye-Sensitized Photo-Supercapacitor Using Electrophoretic Deposition and Compression Methods." *Journal of Power Sources* 195 (18): 6225–31. doi:10.1016/j.jpowsour.2010.01.009

Chen, Qing, Xiaoqian Wang, Fang Chen, Ning Zhang, and Mingming Ma. 2019. "Extremely Strong and Tough Polythiophene Composite for Flexible Electronics." *Chemical Engineering Journal* 368(February): 933–40. doi:10.1016/j.cej.2019.02.203

Chen, Shen Ming, Rasu Ramachandran, Veerappan Mani, and Ramiah Saraswathi. 2014. "Recent Advancements in Electrode Materials for the High-Performance Electrochemical Supercapacitors: A Review." *International Journal of Electrochemical Science* 9 (8): 4072–85.

Chen, Sheng, Junwu Zhu, Xiaodong Wu, Qiaofeng Han, and Xin Wang. 2010. "Graphene Oxide MnO 2." *ACS Nano* 4 (5): 2822–30.

Chen, Tao, and Liming Dai. 2013. "Carbon Nanomaterials for High-Performance Supercapacitors." *Materials Today* 16 (7–8): 272–80. doi:10.1016/j.mattod.2013.07.002

Chen, Wenyong, Xuemei Zhang, Li E. Mo, Yongsheng Zhang, Shuanghong Chen, Xianxi Zhang, and Linhua Hu. 2020. "NiCo2S4 Quantum Dots with High Redox Reactivity for Hybrid Supercapacitors." *Chemical Engineering Journal* 388 (May): 124109. doi:10.1016/J.CEJ.2020.124109

Chen, Xing, Siliang Wang, Junjie Shi, Xiaoyu Du, Qinghua Cheng, Rui Xue, Qiang Wang, Min Wang, Limin Ruan, and Wei Zeng. 2019. "Direct Laser Etching Free-Standing MXene-MoS2 Film for Highly Flexible Micro-Supercapacitor." *Advanced Materials Interfaces* 6 (22): 2–9. doi:10.1002/admi.201901160

Cheng, Qian, Jie Tang, Jun Ma, Han Zhang, Norio Shinya, and Lu Chang Qin. 2011a. "Graphene and Carbon Nanotube Composite Electrodes for Supercapacitors with Ultra-High Energy Density." *Physical Chemistry Chemical Physics* 13 (39): 17615–624. doi:10.1039/c1cp21910c

Cheng, Qian, Jie Tang, Jun Ma, Han Zhang, Norio Shinya, and Lu Chang Qin. 2011b. "Graphene and Nanostructured MnO2 Composite Electrodes for Supercapacitors." *Carbon* 49 (9): 2917–25. doi:10.1016/j.carbon.2011.02.068

Cheng, Tao, You Wei Wu, Ya Li Chen, Yi Zhou Zhang, Wen Yong Lai, and Wei Huang. 2019. "Inkjet-Printed High-Performance Flexible Micro-Supercapacitors with Porous Nanofiber-Like Electrode Structures." *Small* 15 (34): 1–9. doi:10.1002/smll.201901830

Cheng, Xunliang, Jian Pan, Yang Zhao, Meng Liao, and Huisheng Peng. 2018. "Gel Polymer Electrolytes for Electrochemical Energy Storage." *Advanced Energy Materials* 8 (7): 1–16. doi:10.1002/aenm.201702184

Chih, Jui Kung, Anif Jamaluddin, Fuming Chen, Jeng Kuei Chang, and Ching Yuan Su. 2019. "High Energy Density of All-Screen-Printable Solid-State Microsupercapacitors Integrated by Graphene/CNTs as Hierarchical Electrodes." *Journal of Materials Chemistry A* 7 (20): 12779–789. doi:10.1039/c9ta01460h

Chmiola, J., G. Yushin, Y. Gogotsi, C. Portet, P. Simon, P.L. Taberna. n.d. "NEWS OF THE WEEK New ' Supercapacitor ' Promises to Pack More Electrical Punch." *Science*. Author, ROBERT F. SERVICE, year, 2006, page, 902, issue, 5789.

Choi, Changsoon, Jae Ah Lee, A. Young Choi, Youn Tae Kim, Xavier Leprõ, Marcio D. Lima, Ray H. Baughman, and Seon Jeong Kim. 2014. "Flexible Supercapacitor Made of Carbon Nanotube Yarn with Internal Pores." *Advanced Materials* 26 (13): 2059–65. doi:10.1002/adma.201304736

Choi, Hojin, and Hyeonseok Yoon. 2015. "Nanostructured Electrode Materials for Electrochemical Capacitor Applications." *Nanomaterials* 5 (2): 906–36. doi:10.3390/nano5020906

Conway, B.E. 1991a. "Transition from 'supercapacitor' to 'Battery' Behavior in Electrochemical Energy Storage." *Proceedings of the International Power Sources Symposium* 138 (6): 319–27. doi:10.1149/1.2085829

Conway, B.E. 1991b. "Transition from 'supercapacitor' to 'Battery' Behavior in Electrochemical Energy Storage." *Proceedings of the International Power Sources Symposium*, 319–27. doi:10.1149/1.2085829

Conway, B.E. 1995. "Electrochemical Oxide Film Formation at Noble Metals as a Surface-Chemical Process." *Progress in Surface Science* 49 (4): 331–452. doi:10.1016/0079-6816(95)00040-6

Conway, B.E., and W.G. Pell. 2003. "Double-Layer and Pseudocapacitance Types of Electrochemical Capacitors and Their Applications to the Development of Hybrid Devices." *Journal of Solid State Electrochemistry* 7 (9): 637–44. doi:10.1007/s10008-003-0395-7

Conway, Brian E., and Gu Ping. 1991. "Evaluation of Cl• Adsorption in Anodic Cl2 Evolution at Pt by Means of Impedance and Potential-Relaxation Experiments. Influence of the State of Surface Oxidation of the Pt." *Journal of the Chemical Society, Faraday Transactions* 87 (17): 2705–14. doi:10.1039/FT9918702705

Das, Pratteek, Xiaoyu Shi, Qiang Fu, and Zhong Shuai Wu. 2020. "Substrate-Free and Shapeless Planar Micro-Supercapacitors." *Advanced Functional Materials* 30 (7): 1–10. doi:10.1002/adfm.201908758

De, Bibekananda, Soma Banerjee, Tanvi Pal, Kapil Dev Verma, P.K. Manna, and Kamal K. Kar. 2020. "Graphene/Reduced Graphene Oxide as Electrode Materials for Supercapacitors." In *Handbook of Nanocomposite Supercapacitor Materials II*, 302: 271–96. doi:10.1007/978-3-030-52359-6_11

Deng, Jue, Ye Zhang, Yang Zhao, Peining Chen, Xunliang Cheng, and Huisheng Peng. 2015. "A Shape-Memory Supercapacitor Fiber." *Angewandte Chemie - International Edition* 54 (51): 15419–423. doi:10.1002/anie.201508293

Deng, Yalei, Yajun Ji, Fei Chen, Fuyong Ren, and Shufen Tan. 2020. "Superior Performance of Flexible Solid-State Supercapacitors Enabled by Ultrafine Graphene Quantum Dot-Decorated Porous Carbon Spheres." *New Journal of Chemistry* 44 (32): 13591–597. doi:10.1039/D0NJ03163A

Dikin, Dmitriy A., Sasha Stankovich, Eric J. Zimney, Richard D. Piner, Geoffrey H.B. Dommett, Guennadi Evmenenko, Sonbinh T. Nguyen, and Rodney S. Ruoff. 2007. "Preparation and Characterization of Graphene Oxide Paper." *Nature* 448 (7152): 457–60. doi:10.1038/nature06016

Dinh, Ty Mai, Kevin Armstrong, Daniel Guay, and David Pech. 2014. "High-Resolution on-Chip Supercapacitors with Ultra-High Scan Rate Ability." *Journal of Materials Chemistry A* 2 (20): 7170–74. doi:10.1039/c4ta00640b

Doblhofer, Karl, and Chuanjian Zhong. 1991. "The Mechanism of Electrochemical Charge - Transfer Reactions on Conducting Polymer Films." *Synthetic Metals* 43 (1–2): 2865–70. doi:10.1016/0379-6779(91)91191-C

Dong, Liubing, Chengjun Xu, Yang Li, Zheng Hong Huang, Feiyu Kang, Quan Hong Yang, and Xin Zhao. 2016. "Flexible Electrodes and Supercapacitors for Wearable Energy Storage: A Review by Category." *Journal of Materials Chemistry A* 4 (13): 4659–85. doi:10.1039/c5ta10582j

Dong, Liubing, Chengjun Xu, Yang Li, Changle Wu, Baozheng Jiang, Qian Yang, Enlou Zhou, Feiyu Kang, and Quan Hong Yang. 2016. "Simultaneous Production of High-Performance Flexible Textile Electrodes and Fiber Electrodes for Wearable Energy Storage." *Advanced Materials* 28 (8): 1675–81. doi:10.1002/adma.201504747

Du, Chunsheng, and Ning Pan. 2006. "High Power Density Supercapacitor Electrodes of Carbon Nanotube Films by Electrophoretic Deposition." *Nanotechnology* 17 (21): 5314. doi:10.1088/0957-4484/17/21/005

Du, Chunsheng, and Ning Pan. 2007. "Carbon Nanotube-Based Supercapacitors." *Nanotechnology Law and Business* 4 (1): 3–10.

Du, Weimin, Zhiyong Wang, Zhaoqiang Zhu, Sen Hu, Xiaoyan Zhu, Yunfeng Shi, Huan Pang, and Xuefeng Qian. 2014. "Facile Synthesis and Superior Electrochemical Performances of CoNi 2S4/Graphene Nanocomposite Suitable for Supercapacitor Electrodes." *Journal of Materials Chemistry A* 2 (25): 9613–19. doi:10.1039/c4ta00414k

Echendu, O.K., K.B. Okeoma, C.I. Oriaku, and I.M. Dharmadasa. 2016. "Electrochemical Deposition of CdTe Semiconductor Thin Films for Solar Cell Application Using Two-Electrode and Three-Electrode Configurations: A Comparative Study." *Advances in Materials Science and Engineering* 2016: 10–14. doi:10.1155/2016/3581725

El-Kady, Maher F., and Richard B. Kaner. 2013. "Scalable Fabrication of High-Power Graphene Micro-Supercapacitors for Flexible and on-Chip Energy Storage." *Nature Communications* 4: 1–9. doi:10.1038/ncomms2446

Elgrishi, Noémie, Kelley J. Rountree, Brian D. McCarthy, Eric S. Rountree, Thomas T. Eisenhart, and Jillian L. Dempsey. 2018. "A Practical Beginner's Guide to Cyclic Voltammetry." *Journal of Chemical Education* 95 (2): 197–206. doi:10.1021/acs.jchemed.7b00361

Feng, Guang, Jingsong Huang, Bobby G. Sumpter, Vincent Meunier, and Rui Qiao. 2011. "A 'Counter-Charge Layer in Generalized Solvents' Framework for Electrical Double Layers in Neat and Hybrid Ionic Liquid Electrolytes." *Physical Chemistry Chemical Physics* 13 (32): 14723–734. doi:10.1039/c1cp21428d

Feng, Guang, Rui Qiao, and Peter T. Cummings. 2015. "Modeling of Supercapacitors" *Encyclopedia of Microfluidics and Nanofluidics* 2282–2289. doi:10.1007/978-1-4614-5491-5

Ferraro, Dario, Gian Marcello Andolina, Michele Campisi, Vittorio Pellegrini, and Marco Polini. 2019. "Quantum Supercapacitors." *Physical Review B* 100 (7): 075433. doi:10.1103/PHYSREVB.100.075433/FIGURES/4/MEDIUM

Ferraro, Dario, Michele Campisi, Gian Marcello Andolina, Vittorio Pellegrini, and Marco Polini. 2018. "High-Power Collective Charging of a Solid-State Quantum Battery." *Physical Review Letters* 120 (11): 117702. doi:10.1103/PHYSREVLETT.120.117702/FIGURES/3/MEDIUM

Fic, Krzysztof, Elzbieta Frackowiak, and François Béguin. 2012. "Unusual Energy Enhancement in Carbon-Based Electrochemical Capacitors." *Journal of Materials Chemistry* 22 (46): 24213–223. doi:10.1039/c2jm35711a

Foroughi, Javad, Geoffrey M. Spinks, Dennis Antiohos, Azadehsadat Mirabedini, Sanjeev Gambhir, Gordon G. Wallace, Shaban R. Ghorbani, et al. 2014. "Highly Conductive Carbon Nanotube-Graphene Hybrid Yarn." *Advanced Functional Materials* 24 (37): 5859–65. doi:10.1002/adfm.201401412

Forouzandeh, Parnia, Vignesh Kumaravel, and Suresh C. Pillai. 2020. "Electrode Materials for Supercapacitors: A Review of Recent Advances." *Catalysts* 10 (9): 1–73. doi:10.3390/catal10090969

Frackowiak, E., V. Khomenko, K. Jurewicz, K. Lota, and F. Béguin. 2006. "Supercapacitors Based on Conducting Polymers/Nanotubes Composites." *Journal of Power Sources* 153 (2): 413–18. doi:10.1016/j.jpowsour.2005.05.030

Fu, Xudong, Zhangxun Xia, Ruili Sun, Hongyu An, Fulai Qi, Suli Wang, Qingting Liu, and Gongquan Sun. 2018. "A Self-Charging Hybrid Electric Power Device with High Specific Energy and Power." *ACS Energy Letters* 3 (10): 2425–32. doi:10.1021/acsenergylett.8b01331

Fu, Yongping, Xin Cai, Hongwei Wu, Zhibin Lv, Shaocong Hou, Ming Peng, Xiao Yu, and Dechun Zou. 2012. "Fiber Supercapacitors Utilizing Pen Ink for Flexible/Wearable Energy Storage." *Advanced Materials* 24 (42): 5713–18. doi:10.1002/adma.201202930

Fuertes, Antonio B., and Marta Sevilla. 2015. "Superior Capacitive Performance of Hydrochar-Based Porous Carbons in Aqueous Electrolytes." *ChemSusChem* 8 (6): 1049–57. doi:10.1002/cssc.201403267

Gao, Wei, Neelam Singh, Li Song, Zheng Liu, Arava Leela Mohana Reddy, Lijie Ci, Robert Vajtai, Qing Zhang, Bingqing Wei, and Pulickel M. Ajayan. 2011. "Direct Laser Writing of Micro-Supercapacitors on Hydrated Graphite Oxide Films." *Nature Nanotechnology* 6 (8): 496–500. doi:10.1038/nnano.2011.110

Garakani, Mohammad Akbari, Sebastiano Bellani, Vittorio Pellegrini, Reinier Oropesa-Nuñez, Antonio Esau Del Rio Castillo, Sara Abouali, Leyla Najafi, et al. 2021. "Scalable Spray-Coated Graphene-Based Electrodes for High-Power Electrochemical Double-Layer Capacitors Operating over a Wide Range of Temperature." *Energy Storage Materials* 34 (January): 1–11. doi:10.1016/J.ENSM.2020.08.036

Geng, Jiguo, Chuantao Ma, Dong Zhang, and Xuefeng Ning. 2020. "Facile and Fast Synthesis of SnO2 Quantum Dots for High Performance Solid-State Asymmetric Supercapacitor." *Journal of Alloys and Compounds* 825 (June): 153850. doi:10.1016/J.JALLCOM.2020.153850

George, Z. Chen (June 2013). *Progress in Natural Science: Materials International*, June 2013, 23 (3): 245–255.

Ginting, Riski Titian, Manoj Mayaji Ovhal, and Jae Wook Kang. 2018. "A Novel Design of Hybrid Transparent Electrodes for High Performance and Ultra-Flexible Bifunctional Electrochromic-Supercapacitors." *Nano Energy* 53 (May): 650–657. doi:10.1016/j.nanoen.2018.09.016

Giuliani, Gabriele, and Giovanni Vignale. 2005. *Quantum Theory of the Electron Liquid*. Cambridge, UK: Cambridge University Press.

Gnanakan, S. Richard Prabhu, M. Rajasekhar, and A. Subramania. 2009. "Synthesis of Polythiophene Nanoparticles by Surfactant – Assisted Dilute Polymerization Method for High Performance Redox Supercapacitors." *International Journal of Electrochemical Science* 4 (9): 1289–301.

Gogotsi, Yury, and Reginald M. Penner. 2018. "Energy Storage in Nanomaterials – Capacitive, Pseudocapacitive, or Battery-Like?" *ACS Nano* 12 (3): 2081–83. doi:10.1021/acsnano.8b01914

Gouy, M.J.J.P.T.A. 1910. "Sur La Constitution de La Charge Électrique à La Surface d'un Électrolyte." *Hal.Archives-Ouvertes.Fr* 9 (1): 457–468. doi: 10.1051/jphystap:019100090045700ï

Guan, Qun, Jianli Cheng, Bin Wang, Wei Ni, Guifang Gu, Xiaodong Li, Ling Huang, Guangcheng Yang, and Fude Nie. 2014. "Needle-like Co3O4 Anchored on the Graphene with Enhanced Electrochemical Performance for Aqueous Supercapacitors." *ACS Applied Materials and Interfaces* 6 (10): 7626–32. doi:10.1021/am5009369

Gueon, Donghee, and Jun Hyuk Moon. 2015. "Nitrogen-Doped Carbon Nanotube Spherical Particles for Supercapacitor Applications: Emulsion-Assisted Compact Packing and Capacitance Enhancement." *ACS Applied Materials and Interfaces* 7 (36): 20083–089. doi:10.1021/acsami.5b05231

Gujar, T.P., Woo Young Kim, Indra Puspitasari, Kwang Deog Jung, and Oh Shim Joo. 2007. "Electrochemically Deposited Nanograin Ruthenium Oxide as a Pseudocapacitive Electrode." *International Journal of Electrochemical Science* 2 (9): 666–73.

Gulzar, Umair, Subrahmanyam Goriparti, Ermanno Miele, Tao Li, Giulia Maidecchi, Andrea Toma, Francesco De Angelis, Claudio Capiglia, and Remo Proietti Zaccaria. 2016. "Next-Generation Textiles: From Embedded Supercapacitors to Lithium Ion Batteries." *Journal of Materials Chemistry A* 4 (43): 16771–800. doi:10.1039/c6ta06437j

Guo, Kai, Neng Yu, Zhiqiang Hou, Lintong Hu, Ying Ma, Huiqiao Li, and Tianyou Zhai. 2017. "Smart Supercapacitors with Deformable and Healable Functions." *Journal of Materials Chemistry A* 5 (1): 16–30. doi:10.1039/C6TA08458C

Guo, Ruisheng, Jiangtao Chen, Bingjun Yang, Lingyang Liu, Lijun Su, Baoshou Shen, and Xingbin Yan. 2017. "In-Plane Micro-Supercapacitors for an Integrated Device on One Piece of Paper." *Advanced Functional Materials* 27 (43): 1702394–405. doi:10.1002/adfm.201702394

Guo, Yuxuan, Kuaibing Wang, Ye Hong, Hua Wu, and Qichun Zhang. 2021. "Recent Progress on Pristine Two-Dimensional Metal-Organic Frameworks as Active Components in Supercapacitors." *Dalton Transactions* 50 (33): 11331–346. doi:10.1039/d1dt01729b

Hall, Peter J., Mojtaba Mirzaeian, S. Isobel Fletcher, Fiona B. Sillars, Anthony J.R. Rennie, Gbolahan O. Shitta-Bey, Grant Wilson, Andrew Cruden, and Rebecca Carter. 2010. "Energy Storage in Electrochemical Capacitors: Designing Functional Materials to Improve Performance." *Energy and Environmental Science* 3 (9): 1238–51. doi:10.1039/c0ee00004c

Hareesh, K., B. Shateesh, R.P. Joshi, J.F. Williams, D.M. Phase, S.K. Haram, and S.D. Dhole. 2017. "Ultra High Stable Supercapacitance Performance of Conducting Polymer Coated MnO2 Nanorods/RGO Nanocomposites." *RSC Advances* 7 (32): 20027–036. doi:10.1039/c7ra01743j

Harrison, David, Fulian Qiu, John Fyson, Yanmeng Xu, Peter Evans, and Darren Southee. 2013. "A Coaxial Single Fibre Supercapacitor for Energy Storage." *Physical Chemistry Chemical Physics* 15 (29): 12215–219. doi:10.1039/c3cp52036f

Härtel, Andreas, Mathijs Janssen, Daniel Weingarth, Volker Presser, and René Van Roij. 2015. "Heat-to-Current Conversion of Low-Grade Heat from a Thermocapacitive Cycle by Supercapacitors." *Energy and Environmental Science* 8 (8): 2396–401. doi:10.1039/c5ee01192b

Hasan, Mudassir, and Moonyong Lee. 2014. "Enhancement of the Thermo-Mechanical Properties and Efficacy of Mixing Technique in the Preparation of Graphene/PVC Nanocomposites Compared to Carbon Nanotubes/PVC." *Progress in Natural Science: Materials International* 24 (6): 579–87. doi:10.1016/j.pnsc.2014.10.004

He, Haoxuan, Yongming Fu, Tianming Zhao, Xuchao Gao, Lili Xing, Yan Zhang, and Xinyu Xue. 2017. "All-Solid-State Flexible Self-Charging Power Cell Basing on Piezo-Electrolyte for Harvesting/Storing Body-Motion Energy and Powering Wearable Electronics." *Nano Energy* 39 (July): 590–600. doi:10.1016/j.nanoen.2017.07.033

He, Sisi, Longbin Qiu, Lie Wang, Jingyu Cao, Songlin Xie, Qiang Gao, Zhitao Zhang, Jing Zhang, Bingjie Wang, and Huisheng Peng. 2016. "A Three-Dimensionally Stretchable High Performance Supercapacitor." *Journal of Materials Chemistry A* 4 (39): 14968–973. doi:10.1039/C6TA05545A

Heimböckel, Ruben, Frank Hoffmann, and Michael Fröba. 2019. "Insights into the Influence of the Pore Size and Surface Area of Activated Carbons on the Energy Storage of Electric Double Layer Capacitors with a New Potentially Universally Applicable Capacitor Model." *Physical Chemistry Chemical Physics* 21 (6): 3122–33. doi:10.1039/c8cp06443a

Heinrich, Andreas J., William D. Oliver, Lieven M.K. Vandersypen, Arzhang Ardavan, Roberta Sessoli, Daniel Loss, Ania Bleszynski Jayich, Joaquin Fernandez-Rossier, Arne Laucht, and Andrea Morello. 2021. "Quantum-Coherent Nanoscience." *Nature Nanotechnology* 16 (12): 1318–29. doi:10.1038/s41565-021-00994-1

Helmholtz, H. 1879. "Studien Über Electrische Grenzschichten." *Annalen Der Physik* 243 (7): 337–82. doi:10.1002/ANDP.18792430702

Hong, Dajung, and Sanggyu Yim. 2018. "RuO2 Thin Films Electrodeposited on Polystyrene Nanosphere Arrays: Growth Mechanism and Application to Supercapacitor Electrodes." *Langmuir* 34 (14): 4249–54. doi:10.1021/acs.langmuir.8b00829

Hong, Myoung Shin, Seok Hyun Lee, and Sun Wook Kim. 2002. "Use of KCl Aqueous Electrolyte for 2 V Manganese Oxide/Activated Carbon Hybrid Capacitor." *Electrochemical and Solid-State Letters* 5 (10): A227. doi:10.1149/1.1506463/XML

Hong, Xufeng, Liang He, Xinyu Ma, Wei Yang, Yiming Chen, Lei Zhang, Haowu Yan, Zhaohuai Li, and Liqiang Mai. 2017. "Microstructuring of Carbon/Tin Quantum Dots via a Novel Photolithography and Pyrolysis-Reduction Process." *Nano Research* 10 (11): 3743–53. doi:10.1007/s12274-017-1587-2

Hsia, Ben, Julian Marschewski, Shuang Wang, Jung Bin In, Carlo Carraro, Dimos Poulikakos, Costas P. Grigoropoulos, and Roya Maboudian. 2014. "Highly Flexible, All Solid-State Micro-Supercapacitors from Vertically Aligned Carbon Nanotubes." *Nanotechnology* 25 (5): 055401–410. doi:10.1088/0957-4484/25/5/055401

Hsu, Chih Yu, Hsin Wei Chen, Kun Mu Lee, Chih Wei Hu, and Kuo Chuan Ho. 2010. "A Dye-Sensitized Photo-Supercapacitor Based on PProDOT-Et2 Thick Films." *Journal of Power Sources* 195 (18): 6232–38. doi:10.1016/j.jpowsour.2009.12.099

Hu, Liangbing, Mauro Pasta, Fabio La Mantia, Lifeng Cui, Sangmoo Jeong, Heather Dawn Deshazer, Jang Wook Choi, Seung Min Han, and Yi Cui. 2010. "Stretchable, Porous, and Conductive Energy Textiles." *Nano Letters* 10 (2): 708–14. doi:10.1021/nl903949m

Hu, Lin, Nan Yan, Qianwang Chen, Ping Zhang, Hao Zhong, Xinrui Zheng, Yan Li, and Xianyi Hu. 2012. "Fabrication Based on the Kirkendall Effect of Co 3O 4 Porous Nanocages with Extraordinarily High Capacity for Lithium Storage." *Chemistry – A European Journal* 18 (29): 8971–77. doi:10.1002/chem.201200770

Huang, Jingsong, Bobby G. Sumpter, and Vincent Meunier. 2008. "Theoretical Model for Nanoporous Carbon Supercapacitors." *Angewandte Chemie - International Edition* 47 (3): 520–24. doi:10.1002/anie.200703864

Huang, Shifei, Xianglin Zhu, Samrat Sarkar, and Yufeng Zhao. 2019. "Challenges and Opportunities for Supercapacitors." *APL Materials* 7: 100901. doi:10.1063/1.5116146

Huang, Shuo, Fang Wan, Songshan Bi, Jiacai Zhu, Zhiqiang Niu, and Jun Chen. 2019. "A Self-Healing Integrated All-in-One Zinc-Ion Battery." *Angewandte Chemie – International Edition* 58 (13): 4313–17. doi:10.1002/anie.201814653

Huang, Xianwu, and Peiyi Wu. 2020. "A Facile, High-Yield, and Freeze-and-Thaw-Assisted Approach to Fabricate MXene with Plentiful Wrinkles and Its Application in On-Chip Micro-Supercapacitors." *Advanced Functional Materials* 30 (12): 1–11. doi:10.1002/adfm.201910048

Huang, Yan, Hong Hu, Yang Huang, Minshen Zhu, Wenjun Meng, Chang Liu, Zengxia Pei, Chonglei Hao, Zuankai Wang, and Chunyi Zhi. 2015. "From Industrially Weavable and Knittable Highly Conductive Yarns to Large Wearable Energy Storage Textiles." *ACS Nano* 9 (5): 4766–75. doi:10.1021/acsnano.5b00860

Huang, Yan, Ming Zhong, Yang Huang, Minshen Zhu, Zengxia Pei, Zifeng Wang, Qi Xue, Xuming Xie, and Chunyi Zhi. 2015. "A Self-Healable and Highly Stretchable Supercapacitor Based on a Dual Crosslinked Polyelectrolyte." *Nature Communications* 6: 10310–18. doi:10.1038/ncomms10310

Huo, Pengfei, Shoupeng Ni, Pu Hou, Zhiyu Xun, Yang Liu, and Jiyou Gu. 2019. "A Crosslinked Soybean Protein Isolate Gel Polymer Electrolyte Based on Neutral Aqueous Electrolyte for a High-Energy-Density Supercapacitor." *Polymers* 11 (5): 863–875. doi:10.3390/polym11050863

Ilyanok, Alexandr Mikhailovich. 2006. Quantum supercapacitor. US7193261B2, issued July 1, 2006.

In, Jung Bin, Ben Hsia, Jae Hyuck Yoo, Seungmin Hyun, Carlo Carraro, Roya Maboudian, and Costas P. Grigoropoulos. 2015. "Facile Fabrication of Flexible All Solid-State Micro-Supercapacitor by Direct Laser Writing of Porous Carbon in Polyimide." *Carbon* 83: 144–51. doi:10.1016/j.carbon.2014.11.017

Iro, Zaharaddeen S., C. Subramani, and S.S. Dash. 2016. "A Brief Review on Electrode Materials for Supercapacitor." *International Journal of Electrochemical Science* 11 (12): 10628–643. doi:10.20964/2016.12.50

Jadhav, Vijaykumar V., Rajaram S. Mane, and Pritamkumar V. Shinde. 2020. "Electrochemical Supercapacitors: History, Types, Designing Processes, Operation Mechanisms, and Advantages and Disadvantages." In *Bismuth-Ferrite-Based Electrochemical Supercapacitors*, 11–36. Springer, Cham. doi:10.1007/978-3-030-16718-9_2

Jaidev, and S. Ramaprabhu. 2012. "Poly(p-Phenylenediamine)/Graphene Nanocomposites for Supercapacitor Applications." *Journal of Materials Chemistry* 22 (36): 18775–783. doi:10.1039/c2jm33627h

Jayalakshmi, M., and K. Balasubramanian. 2008a. “Simple Capacitors to Supercapacitors – An Overview.” *International Journal of Electrochemical Science* 3 (11): 1196–217.

Jayalakshmi, M., and K. Balasubramanian. 2008b. “Simple Capacitors to Supercapacitors – An Overview.” *International Journal of Electrochemical Science* 3 (11): 1196–217.

Jha, Mihir Kumar, Binson Babu, Bradyn J. Parker, Vishnu Surendran, Neil R. Cameron, Manikoth M. Shaijumon, and Chandramouli Subramaniam. 2020. “Hierarchically Engineered Nanocarbon Florets as Bifunctional Electrode Materials for Adsorptive and Intercalative Energy Storage.” *ACS Applied Materials and Interfaces* 12 (38): 42669–677. doi:10.1021/acsami.0c09021

Jha, Mihir Kumar, Ranadeb Ball, Raghunandan Seelaboyina, and Chandramouli Subramaniam. 2020. “All Solid-State Coaxial Supercapacitor with Ultrahigh Scan Rate Operability of 250»000 MV/s by Thermal Engineering of the Electrode-Electrolyte Interface.” *ACS Applied Energy Materials* 3 (4): 3454–64. doi:10.1021/acsaem.9b02528

Jha, Mihir Kumar, Kenji Hata, and Chandramouli Subramaniam. 2019. “Interwoven Carbon Nanotube Wires for High-Performing, Mechanically Robust, Washable, and Wearable Supercapacitors.” *ACS Applied Materials and Interfaces* 11 (20): 18285–294. doi:10.1021/acsami.8b22233

Jha, Mihir Kumar, Tanya Jain, and Chandramouli Subramaniam. 2020. “Origami of Solid-State Supercapacitive Microjunctions Operable at 3 v with High Specific Energy Density for Wearable Electronics.” *ACS Applied Electronic Materials* 2 (3): 659–69. doi:10.1021/acsaelm.9b00769

Ji, Hengxing, Xin Zhao, Zhenhua Qiao, Jeil Jung, Yanwu Zhu, Yalin Lu, Li Li Zhang, Allan H. MacDonald, and Rodney S. Ruoff. 2014. “Capacitance of Carbon-Based Electrical Double-Layer Capacitors.” *Nature Communications* 5 (Cmcm): 3317–23. article no. 3317. doi:10.1038/ncomms4317

Jiang, Kaiyue, Igor A. Baburin, Peng Han, Chongqing Yang, Xiaobin Fu, Yefeng Yao, Jiantong Li, et al. 2020. “Interfacial Approach toward Benzene-Bridged Polypyrrole Film–Based Micro-Supercapacitors with Ultrahigh Volumetric Power Density.” *Advanced Functional Materials* 30 (7): 1–9. doi:10.1002/adfm.201908243

Jiang, Qiu, Changsheng Wu, Zhengjun Wang, Aurelia Chi Wang, Jr Hau He, Zhong Lin Wang, and Husam N. Alshareef. 2018. “MXene Electrochemical Microsupercapacitor Integrated with Triboelectric Nanogenerator as a Wearable Self-Charging Power Unit.” *Nano Energy* 45 (December 2017): 266–72. doi:10.1016/j.nanoen.2018.01.004

Jiao, Shangqing, Aiguo Zhou, Mingzai Wu, and Haibo Hu. 2019. “Kirigami Patterning of MXene/Bacterial Cellulose Composite Paper for All-Solid-State Stretchable Micro-Supercapacitor Arrays.” *Advanced Science* 6 (12): 1900529–41. doi:10.1002/advs.201900529

Jost, Kristy, Genevieve Dion, and Yury Gogotsi. 2014. “Textile Energy Storage in Perspective.” *Journal of Materials Chemistry A* 2 (28): 10776–87. doi:10.1039/c4ta00203b

Jost, Kristy, David P. Durkin, Luke M. Haverhals, E. Kathryn Brown, Matthew Langenstein, Hugh C. De Long, Paul C. Trulove, Yury Gogotsi, and Genevieve Dion. 2015. “Natural Fiber Welded Electrode Yarns for Knittable Textile Supercapacitors.” *Advanced Energy Materials* 5 (4): 1–8. doi:10.1002/aenm.201401286

Kaiser, Kenneth L. 2006. *Electrostatic Discharge*. Taylor & Francis.

Kakaei, Karim, Mehdi D. Esrafili, and Ali Ehsani. 2019. “Graphene-Based Electrochemical Supercapacitors.” *Interface Science and Technology* 27 (1): 339–86. doi:10.1016/B978-0-12-814523-4.00009-5

Karthika, Prasanna, Natarajan Rajalakshmi, and Kaveripatnam S. Dhathathreyan. 2012. "Functionalized Exfoliated Graphene Oxide as Supercapacitor Electrodes." *Soft Nanoscience Letters* 02 (04): 59–66. doi:10.4236/snl.2012.24011

Kelly-Holmes, Helen. 2016. "Advertising as Multilingual Communication." *Advertising as Multilingual Communication* 45: 1–206. doi:10.1057/9780230503014

Keskinen, Jari, Saara Tuurala, Martin Sjödin, Kaisa Kiri, Leif Nyholm, Timo Flyktman, Maria Strømme, and Maria Smolander. 2015. "Asymmetric and Symmetric Supercapacitors Based on Polypyrrole and Activated Carbon Electrodes." *Synthetic Metals* 203: 192–99. doi:10.1016/j.synthmet.2015.02.034

Kiamahalleh, Meisam Valizadeh, Sharif Hussein Sharif Zein, Ghasem Najafpour, Suhairi Abd Sata, and Surani Buniran. 2012. "Multiwalled Carbon Nanotubes Based Nanocomposites for Supercapacitors: A Review of Electrode Materials." *Nano* 7 (2): 1–27. doi:10.1142/S1793292012300022

Kim, Hansung, and Branko N. Popov. 2003. "Synthesis and Characterization of MnO2-Based Mixed Oxides as Supercapacitors." *Journal of The Electrochemical Society* 150 (3): D56. doi:10.1149/1.1541675/XML

Kim, Suk Lae, Henry Taisun Lin, and Choongho Yu. 2016. "Thermally Chargeable Solid-State Supercapacitor." *Advanced Energy Materials* 6 (18): 1–7. doi:10.1002/aenm.201600546

Kim, Taeyoung, Gyujin Jung, Seonmi Yoo, Kwang S. Suh, and Rodney S. Ruoff. 2013. "Activated Graphene-Based Carbons as Supercapacitor Electrodes with Macro- and Mesopores." *ACS Nano* 7 (8): 6899–905. doi:10.1021/nn402077v

Krishnamoorthy, Karthikeyan, Parthiban Pazhamalai, Vimal Kumar Mariappan, Swapnil Shital Nardekar, Surjit Sahoo, and Sang Jae Kim. 2020. "Probing the Energy Conversion Process in Piezoelectric-Driven Electrochemical Self-Charging Supercapacitor Power Cell Using Piezoelectrochemical Spectroscopy." *Nature Communications* 11 (1): 1–11. doi:10.1038/s41467-020-15808-6

Kumar, Sumana, and Abha Misra. 2021. "Three-Dimensional Carbon Foam-Metal Oxide-Based Asymmetric Electrodes for High-Performance Solid-State Micro-Supercapacitors." *Nanoscale* 13 (46): 19453–65. doi:10.1039/d1nr02833b

Kumar, Sumana, Swanand Telpande, Veera Manikandan, Praveen Kumar, and Abha Misra. 2020. "Novel Electrode Geometry for High Performance CF/Fe2O3based Planar Solid State Micro-Electrochemical Capacitors." *Nanoscale* 12 (37): 19438–49. doi:10.1039/d0nr04410e

Kurra, Narendra, M.K. Hota, and H.N. Alshareef. 2015. "Conducting Polymer Micro-Supercapacitors for Flexible Energy Storage and Ac Line-Filtering." *Nano Energy* 13: 500–08. doi:10.1016/j.nanoen.2015.03.018

Laforgue, Alexis, Patrice Simon, Christian Sarrazin, and Jean François Fauvarque. 1999. "Polythiophene-Based Supercapacitors." *Journal of Power Sources* 80 (1): 142–48. doi:10.1016/S0378-7753(98)00258-4

Lahrar, El Hassane, Patrice Simon, and Céline Merlet. 2021. "Carbon–Carbon Supercapacitors: Beyond the Average Pore Size or How Electrolyte Confinement and Inaccessible Pores Affect the Capacitance." *The Journal of Chemical Physics* 155 (18): 184703. doi:10.1063/5.0065150

Laszczyk, Karolina U., Kazufumi Kobashi, Shunsuke Sakurai, Atsuko Sekiguchi, Don N. Futaba, Takeo Yamada, and Kenji Hata. 2015. "Lithographically Integrated Microsupercapacitors for Compact, High Performance, and Designable Energy Circuits." *Advanced Energy Materials* 5 (18): 6–11. doi:10.1002/aenm.201500741

Lee, Gwan Hyoung, Chul Ho Lee, Arend M. Van Der Zande, Minyong Han, Xu Cui, Ghidewon Arefe, Colin Nuckolls, Tony F. Heinz, James Hone, and Philip Kim. 2014. "Heterostructures Based on Inorganic and Organic van Der Waals Systems." *APL Materials* 2 (9): 092511. doi:10.1063/1.4894435

Lee, Jae Kyung, Habib M. Pathan, Kwang Deog Jung, and Oh Shim Joo. 2006. "Electrochemical Capacitance of Nanocomposite Films Formed by Loading Carbon Nanotubes with Ruthenium Oxide." *Journal of Power Sources* 159 (2): 1527–31. doi:10.1016/j.jpowsour.2005.11.063

Lee, Sang Young, Keun Ho Choi, Woo Sung Choi, Yo Han Kwon, Hye Ran Jung, Heon Cheol Shin, and Je Young Kim. 2013. "Progress in Flexible Energy Storage and Conversion Systems, with a Focus on Cable-Type Lithium-Ion Batteries." *Energy and Environmental Science* 6 (8): 2414–23. doi:10.1039/c3ee24260a

Lee, Ying Hui, Kuo Hsin Chang, and Chi Chang Hu. 2013. "Differentiate the Pseudocapacitance and Double-Layer Capacitance Contributions for Nitrogen-Doped Reduced Graphene Oxide in Acidic and Alkaline Electrolytes." *Journal of Power Sources* 227: 300–08. doi:10.1016/j.jpowsour.2012.11.026

Li, Bing, Kosuke Igawa, Jianwei Chai, Ye Chen, Yong Wang, Derrick Wenhui Fam, Nguk Neng Tham, et al. 2020. "String of Pyrolyzed ZIF-67 Particles on Carbon Fibers for High-Performance Electrocatalysis." *Energy Storage Materials* 25 (April 2019): 137–44. doi:10.1016/j.ensm.2019.10.021

Li, Hanlu, Jixiao Wang, Qingxian Chu, Zhi Wang, Fengbao Zhang, and Shichang Wang. 2009. "Theoretical and Experimental Specific Capacitance of Polyaniline in Sulfuric Acid." *Journal of Power Sources* 190 (2): 578–86. doi:10.1016/j.jpowsour.2009.01.052

Li, Hong, Zhaoxiang Wang, Liquan Chen, and Xuejie Huang. 2009. "Research on Advanced Materials for Li-Ion Batteries." *Advanced Materials* 21 (45): 4593–607. doi:10.1002/adma.200901710

Li, Hongpeng, Xiran Li, Jiajie Liang, and Yongsheng Chen. 2019. "Hydrous RuO 2 -Decorated MXene Coordinating with Silver Nanowire Inks Enabling Fully Printed Micro-Supercapacitors with Extraordinary Volumetric Performance." *Advanced Energy Materials* 9 (15): 1–13. doi:10.1002/aenm.201803987

Li, Huili, Tian Lv, Huanhuan Sun, Guiju Qian, Ning Li, Yao Yao, and Tao Chen. 2019. "Ultrastretchable and Superior Healable Supercapacitors Based on a Double Cross-Linked Hydrogel Electrolyte." *Nature Communications* 10 (1): 1–8. doi:10.1038/s41467-019-08320-z

Li, Jiantong, Szymon Sollami Delekta, Panpan Zhang, Sheng Yang, Martin R. Lohe, Xiaodong Zhuang, Xinliang Feng, and Mikael Östling. 2017. "Scalable Fabrication and Integration of Graphene Microsupercapacitors through Full Inkjet Printing." *ACS Nano* 11 (8): 8249–56. doi:10.1021/acsnano.7b03354

Li, Lei, Jibo Zhang, Zhiwei Peng, Yilun Li, Caitian Gao, Yongsung Ji, Ruquan Ye, et al. 2016. "High-Performance Pseudocapacitive Microsupercapacitors from Laser-Induced Graphene." *Advanced Materials* 28 (5): 838–45. doi:10.1002/adma.201503333

Li, Mei, and Lanlan Yang. 2015. "Intrinsic Flexible Polypyrrole Film with Excellent Electrochemical Performance." *Journal of Materials Science: Materials in Electronics* 26 (7): 4875–79. doi:10.1007/s10854-015-2996-1

Li, Tianyu, Huiying Liu, Dingxuan Zhao, and Lili Wang. 2016. "Design and Analysis of a Fuel Cell Supercapacitor Hybrid Construction Vehicle." *International Journal of Hydrogen Energy* 41 (28): 12307–19. doi:10.1016/j.ijhydene.2016.05.040

Li, Weiwei, Azat Meredov, and Atif Shamim. 2019. "Coat-and-Print Patterning of Silver Nanowires for Flexible and Transparent Electronics." *Npj Flexible Electronics* 3 (1): 1–7. doi: 10.1038/s41528-019-0063-3

Li, Yang, Huaqing Xie, Jing Li, Yoshio Bando, Yusuke Yamauchi, and Joel Henzie. 2019. "High Performance Nanoporous Carbon Microsupercapacitors Generated by a Solvent-Free MOF-CVD Method." *Carbon* 152: 688–96. doi:10.1016/j.carbon.2019.06.050

Li, Yingzhi, Xin Zhao, Pingping Yu, and Qinghua Zhang. 2013. "Oriented Arrays of Polyaniline Nanorods Grown on Graphite Nanosheets for an Electrochemical Supercapacitor." *Langmuir* 29 (1): 493–500. doi:10.1021/la303632d

Lian, Yue, Zongying Xu, Dawei Wang, Yongqing Bai, Chaolei Ban, Jing Zhao, and Huaihao Zhang. 2021. "Nb2O5 Quantum Dots Coated with Biomass Carbon for Ultra-Stable Lithium-Ion Supercapacitors." *Journal of Alloys and Compounds* 850 (January): 156808. doi:10.1016/J.JALLCOM.2020.156808

Lin, Rongying, Pierre-Louis Taberna, Sébastien Fantini, Volker Presser, Carlos R. Pérez, François Malbosc, Nalin L. Rupesinghe, Kenneth B.K. Teo, Yury Gogotsi, and Patrice Simon. 2011. "Capacitive Energy Storage from −50 to 100 °C Using an Ionic Liquid Electrolyte." *The Journal of Physical Chemistry Letters* 2 (19): 2396–401. doi:10.1021/jz201065t

Lin, Yuanjing, Yuan Gao, and Zhiyong Fan. 2017. "Printable Fabrication of Nanocoral-Structured Electrodes for High-Performance Flexible and Planar Supercapacitor with Artistic Design." *Advanced Materials* 29 (43): 1–8. doi:10.1002/adma.201701736

Lindström, Henrik, Sven Södergren, Anita Solbrand, Håkan Rensmo, Johan Hjelm, Anders Hagfeldt, and Sten Eric Lindquist. 1997. "Li+ Ion Insertion in TiO2 (Anatase). 1. Chronoamperometry on CVD Films and Nanoporous Films." *Journal of Physical Chemistry B* 101 (39): 7710–16. doi:10.1021/jp970489r

Liu, Chenguang, Zhenning Yu, David Neff, Aruna Zhamu, and Bor Z. Jang. 2010. "Graphene-Based Supercapacitor with an Ultrahigh Energy Density." *Nano Letters* 10 (12): 4863–68. doi:10.1021/nl102661q

Liu, Fei, Shuyan Song, Dongfeng Xue, and Hongjie Zhang. 2012. "Selective Crystallization with Preferred Lithium-Ion Storage Capability of Inorganic Materials." *Nanoscale Research Letters* 7 (1): 1–17. doi:10.1186/1556-276X-7-149

Liu, Jilei, Jin Wang, Chaohe Xu, Hao Jiang, Chunzhong Li, Lili Zhang, Jianyi Lin, and Ze Xiang Shen. 2018. "Advanced Energy Storage Devices: Basic Principles, Analytical Methods, and Rational Materials Design." *Advanced Science* 5 (1): 1700322–41. doi:10.1002/advs.201700322

Liu, Li, Qiang Lu, Shuanglei Yang, Jiang Guo, Qingyong Tian, Weijing Yao, Zhanhu Guo, Vellaisamy A.L. Roy, and Wei Wu. 2018. "All-Printed Solid-State Microsupercapacitors Derived from Self-Template Synthesis of Ag@PPy Nanocomposites." *Advanced Materials Technologies* 3 (1): 1–9. doi:10.1002/admt.201700206

Liu, Lingyang, Baoshou Shen, Dan Jiang, Ruisheng Guo, Lingbin Kong, and Xingbin Yan. 2016. "Watchband-Like Supercapacitors with Body Temperature Inducible Shape Memory Ability." *Advanced Energy Materials* 6 (16): 1–10. doi:10.1002/aenm.201600763

Liu, Panbo, Jing Yan, Zhaoxu Guang, Ying Huang, Xifei Li, and Wenhuan Huang. 2019. "Recent Advancements of Polyaniline-Based Nanocomposites for Supercapacitors." *Journal of Power Sources* 424 (March): 108–30. doi:10.1016/j.jpowsour.2019.03.094

Liu, Peng, Xue Wang, and Yunjiao Wang. 2014. "Design of Carbon Black/Polypyrrole Composite Hollow Nanospheres and Performance Evaluation as Electrode Materials for Supercapacitors." *ACS Sustainable Chemistry and Engineering* 2 (7): 1795–801. doi:10.1021/sc5001034

Liu, Wen Wen, Ya Qiang Feng, Xing Bin Yan, Jiang Tao Chen, and Qun Ji Xue. 2013. "Superior Micro-Supercapacitors Based on Graphene Quantum Dots." *Advanced Functional Materials* 23 (33): 4111–22. doi:10.1002/adfm.201203771

Liu, Y., S.P. Jiang, and Z. Shao. 2020. "Intercalation Pseudocapacitance in Electrochemical Energy Storage: Recent Advances in Fundamental Understanding and Materials Development." *Materials Today Advances* 7: 100072. doi:10.1016/j.mtadv.2020.100072

Liu, Yu, Baihe Zhang, Yaqiong Yang, Zheng Chang, Zubiao Wen, and Yuping Wu. 2013. "Polypyrrole-Coated α-MoO3 Nanobelts with Good Electrochemical Performance as Anode Materials for Aqueous Supercapacitors." *Journal of Materials Chemistry A* 1 (43): 13582–87. doi:10.1039/c3ta12902k

Lota, Katarzyna, Agnieszka Sierczynska, Ilona Acznik, and Grzegorz Lota. 2013. "Effect of Aqueous Electrolytes on Electrochemical Capacitor Capacitance." *Chemik* 67 (11): 1138–45.

Lu, An Kang, Han Yu Li, and Yao Yu. 2019. "Holey Graphene Synthesized by Electrochemical Exfoliation for High-Performance Flexible Microsupercapacitors." *Journal of Materials Chemistry A* 7 (13): 7852–58. doi:10.1039/C9TA00792J

Lu, Ke, Dan Li, Xiang Gao, Hongxiu Dai, Nan Wang, and Houyi Ma. 2015. "An Advanced Aqueous Sodium-Ion Supercapacitor with a Manganous Hexacyanoferrate Cathode and a Fe3O4/RGO Anode." *Journal of Materials Chemistry A* 3 (31): 16013–19. doi:10.1039/c5ta04244e

Lufrano, F., and P. Staiti. 2010. "Mesoporous Carbon Materials as Electrodes for Electrochemical Supercapacitors." *International Journal of Electrochemical Science* 5 (6): 903–16.

Lukatskaya, Maria R., Olha Mashtalir, Chang E. Ren, Yohan Dall'Agnese, Patrick Rozier, Pierre Louis Taberna, Michael Naguib, Patrice Simon, Michel W. Barsoum, and Yury Gogotsi. 2013. "Cation Intercalation and High Volumetric Capacitance of Two-Dimensional Titanium Carbide." *Science* 341 (6153): 1502–05. doi:10.1126/science.1241488

Lv, Wei, Dai Ming Tang, Yan Bing He, Cong Hui You, Zhi Qiang Shi, Xue Cheng Chen, Cheng Meng Chen, Peng Xiang Hou, Chang Liu, and Quan Hong Yang. 2009. "Low-Temperature Exfoliated Graphenes: Vacuum-Promoted Exfoliation and Electrochemical Energy Storage." *ACS Nano* 3 (11): 3730–36. doi:10.1021/nn900933u

Mane, Rajaram S., Jinho Chang, Dukho Ham, B.N. Pawar, T. Ganesh, Byung Won Cho, Joon Kee Lee, and Sung Hwan Han. 2009. "Dye-Sensitized Solar Cell and Electrochemical Supercapacitor Applications of Electrochemically Deposited Hydrophilic and Nanocrystalline Tin Oxide Film Electrodes." *Current Applied Physics* 9 (1): 87–91. doi:10.1016/j.cap.2007.11.013

Mao, Lu, Kai Zhang, Hardy Sze On Chan, and Jishan Wu. 2012. "Surfactant-Stabilized Graphene/Polyaniline Nanofiber Composites for High Performance Supercapacitor Electrode." *Journal of Materials Chemistry* 22 (1): 80–85. doi:10.1039/c1jm12869h

Mathew, Elma Elizaba, and Manoj Balachandran. 2021. "Crumpled and Porous Graphene for Supercapacitor Applications: A Short Review." *Carbon Letters* 31 (4): 537–55. doi:10.1007/s42823-021-00229-2

Mathis, Tyler S., Narendra Kurra, Xuehang Wang, David Pinto, Patrice Simon, and Yury Gogotsi. 2019. "Energy Storage Data Reporting in Perspective—Guidelines for Interpreting the Performance of Electrochemical Energy Storage Systems." *Advanced Energy Materials* 9 (39): 1902007–20. doi:10.1002/aenm.201902007

Meng, Yuning, Yang Zhao, Chuangang Hu, Huhu Cheng, Yue Hu, Zhipan Zhang, Gaoquan Shi, and Liangti Qu. 2013. "All-Graphene Core-Sheath Microfibers for All-Solid-State, Stretchable Fibriform Supercapacitors and Wearable Electronic Textiles." *Advanced Materials* 25 (16): 2326–31. doi:10.1002/adma.201300132

Merlet, Céline, Benjamin Rotenberg, Paul A. Madden, Pierre Louis Taberna, Patrice Simon, Yury Gogotsi, and Mathieu Salanne. 2012. "On the Molecular Origin of Supercapacitance in Nanoporous Carbon Electrodes." *Nature Materials* 11 (4): 306–10. doi:10.1038/nmat3260

Mishra, Rajneesh Kumar, Gyu Jin Choi, Youngku Sohn, Seung Hee Lee, and Jin Seog Gwag. 2019. "Reduced Graphene Oxide Based Supercapacitors: Study of Self-Discharge Mechanisms, Leakage Current and Stability via Voltage Holding Tests." *Materials Letters* 253 (October): 250–54. doi:10.1016/J.MATLET.2019.06.073

Mohapatra, S., A. Acharya, and G.S. Roy. 2012. "The Role of Nanomaterial for the Design of Supercapacitors." *Latin-American Journal of Physics Education* 6 (3): 380–84.

Mondal, Sudeshna, and Chandramouli Subramaniam. 2019. "Point-of-Care, Cable-Type Electrochemical Zn2+ Sensor with Ultrahigh Sensitivity and Wide Detection Range for Soil and Sweat Analysis." *ACS Sustainable Chemistry and Engineering* 7 (17): 14569–79. doi:10.1021/acssuschemeng.9b02173

Murakami, Takurou N., Norimichi Kawashima, and Tsutomu Miyasaka. 2005. "A High-Voltage Dye-Sensitized Photocapacitor of a Three-Electrode System." *Chemical Communications*, no. 26: 3346–48. doi:10.1039/b503122b

Naik, Smita Gajanan, and Mohammad Hussain K. Rabinal. 2020. "Molybdenum Disulphide Heterointerfaces as Potential Materials for Solar Cells, Energy Storage, and Hydrogen Evolution." *Energy Technology* 8 (6): 1901299. doi:10.1002/ENTE.201901299

Najib, Sumaiyah, and Emre Erdem. 2019. "Current Progress Achieved in Novel Materials for Supercapacitor Electrodes: Mini Review." *Nanoscale Advances* 1 (8): 2817–27. doi:10.1039/c9na00345b

Nam, Min Sik, Umakant Patil, Byeongho Park, Heung Bo Sim, and Seong Chan Jun. 2016. "A Binder Free Synthesis of 1D PANI and 2D MoS2 Nanostructured Hybrid Composite Electrodes by the Electrophoretic Deposition (EPD) Method for Supercapacitor Application." *RSC Advances* 6 (103): 101592–601. doi:10.1039/c6ra16078f

Namsheer, K., and Chandra Sekhar Rout. 2021. "Photo-Powered Integrated Supercapacitors: A Review on Recent Developments, Challenges and Future Perspectives." *Journal of Materials Chemistry A* 9 (13): 8248–78. doi:10.1039/D1TA00444A

Naoi, Katsuhiko, and Patrice Simon. 2008. "New Materials and New Confgurations for Advanced Electrochemical Capacitors." *Electrochemical Society Interface* 17 (1): 34–37. doi:10.1149/2.f04081if

Naskar, Pappu, Apurba Maiti, Priyanka Chakraborty, Debojyoti Kundu, Biplab Biswas, and Anjan Banerjee. 2021. "Chemical Supercapacitors: A Review Focusing on Metallic Compounds and Conducting Polymers." *Journal of Materials Chemistry A* 9 (4): 1970–2017. doi:10.1039/d0ta09655e

Nasrin, Kabeer, Subramaniam Gokulnath, Manickavasakam Karnan, Kaipannan Subramani, and Marappan Sathish. 2021. "Redox-Additives in Aqueous, Non-Aqueous, and All-Solid-State Electrolytes for Carbon-Based Supercapacitor: A Mini-Review." *Energy and Fuels* 35 (8): 6465–82. doi:10.1021/acs.energyfuels.1c00341

Ngamchuea, Kamonwad, Shaltiel Eloul, Kristina Tschulik, and Richard G. Compton. 2014. "Planar Diffusion to Macro Disc Electrodes—What Electrode Size Is Required for the Cottrell and Randles-Sevcik Equations to Apply Quantitatively?" *Journal of Solid State Electrochemistry* 18 (12): 3251–57. doi:10.1007/s10008-014-2664-z

Niu, Zhiqiang, Haibo Dong, Bowen Zhu, Jinzhu Li, Huey Hoon Hng, Weiya Zhou, Xiaodong Chen, and Sishen Xie. 2013. "Highly Stretchable, Integrated Supercapacitors Based on Single-Walled Carbon Nanotube Films with Continuous Reticulate Architecture." *Advanced Materials* 25 (7): 1058–64. doi:10.1002/adma.201204003

Niu, Zhiqiang, Li Zhang, Lili Liu, Bowen Zhu, Haibo Dong, and Xiaodong Chen. 2013. "All-Solid-State Flexible Ultrathin Micro-Supercapacitors Based on Graphene." *Advanced Materials* 25 (29): 4035–42. doi:10.1002/adma.201301332

Niu, Zhiqiang, Weiya Zhou, Jun Chen, Guoxing Feng, Hong Li, Wenjun Ma, Jinzhu Li, et al. 2011. "Compact-Designed Supercapacitors Using Free-Standing Single-Walled Carbon Nanotube Films." *Energy and Environmental Science* 4 (4): 1440–46. doi:10.1039/c0ee00261e

Noh, Y., & Aluru, N. R. (2020). Ion Transport in Electrically Imperfect Nanopores. *ACS Nano*, 14 (8): 10518–1052610.1021/acsnano.0c04453.

Oh, Young Joon, Jung Joon Yoo, Yong Il Kim, Jae Kook Yoon, Ha Na Yoon, Jong Huy Kim, and Seung Bin Park. 2014. “Oxygen Functional Groups and Electrochemical Capacitive Behavior of Incompletely Reduced Graphene Oxides as a Thin-Film Electrode of Supercapacitor.” *Electrochimica Acta* 116: 118–28. doi:10.1016/j.electacta.2013.11.040

Owusu, Kwadwo Asare, Longbing Qu, Jiantao Li, Zhaoyang Wang, Kangning Zhao, Chao Yang, Kalele Mulonda Hercule, et al. 2017. “Low-Crystalline Iron Oxide Hydroxide Nanoparticle Anode for High-Performance Supercapacitors.” *Nature Communications* 8 (1): 14264–75. doi:10.1038/ncomms14264

Pandolfo, A.G., and A.F. Hollenkamp. 2006. “Carbon Properties and Their Role in Supercapacitors.” *Journal of Power Sources* 157 (1): 11–27. doi:10.1016/j.jpowsour.2006.02.065

Pang, Huan, Yizhou Zhang, Wen Yong Lai, Zheng Hu, and Wei Huang. 2015. “Lamellar K2Co3(P2O7)2·2H2O Nanocrystal Whiskers: High-Performance Flexible All-Solid-State Asymmetric Micro-Supercapacitors via Inkjet Printing.” *Nano Energy* 15: 303–12. doi:10.1016/j.nanoen.2015.04.034

Park, Nam-gyu, Arthur J. Frank, and South Korea. 2003. “Effect of Morphology on Electron Transport in Dye-Sensitized Nanostructured TiO2 Films.” *Journal of Photoscience* 10: 199–202.

Patil, B.H., A.D. Jagadale, and C.D. Lokhande. 2012. “Synthesis of Polythiophene Thin Films by Simple Successive Ionic Layer Adsorption and Reaction (SILAR) Method for Supercapacitor Application.” *Synthetic Metals* 162 (15–16): 1400–05. doi:10.1016/j.synthmet.2012.05.023

Patil, B.H., S.J. Patil, and C.D. Lokhande. 2014. “Electrochemical Characterization of Chemically Synthesized Polythiophene Thin Films: Performance of Asymmetric Supercapacitor Device.” *Electroanalysis* 26 (9): 2023–32. doi:10.1002/elan.201400284

Pazhamalai, Parthiban, Karthikeyan Krishnamoorthy, Vimal Kumar Mariappan, Surjit Sahoo, Sindhuja Manoharan, and Sang Jae Kim. 2018. “A High Efficacy Self-Charging MoSe2 Solid-State Supercapacitor Using Electrospun Nanofibrous Piezoelectric Separator with Ionogel Electrolyte.” *Advanced Materials Interfaces* 5 (12): 1–9. doi:10.1002/admi.201800055

Pean, C., B. Rotenberg, P. Simon, and M. Salanne. 2016. “Multi-Scale Modelling of Supercapacitors: From Molecular Simulations to a Transmission Line Model.” *Journal of Power Sources* 326: 680–85. doi:10.1016/j.jpowsour.2016.03.095

Peng, Xiao , Shuhai, Quan , and Changjun, Xie. 2017. A New Supercapacitor and Li-ion Battery Hybrid System for Electric Vehicle in ADVISOR. *Journal of Physics: Conference Series* 806. 012015–21, doi: 10.1088/1742-6596/806/1/012015

Pech, David, Magali Brunet, Ty Mai Dinh, Kevin Armstrong, Julie Gaudet, and Daniel Guay. 2013. “Influence of the Configuration in Planar Interdigitated Electrochemical Micro-Capacitors.” *Journal of Power Sources* 230: 230–35. doi:10.1016/j.jpowsour.2012.12.039

Pech, David, Magali Brunet, Pierre-louis Taberna, Patrice Simon, Fabien Mesnilgrente, Véronique Conédéra, Hugo Durou, et al. 2017. “Elaboration of a Microstructured Inkjet-Printed Carbon Electrochemical Capacitor To Cite This Version: HAL Id: Hal-01443055.” *Journal of Power Sources*, 15 February 2010, 195 (4): 1266–1269.

Peng, Lele, Xu Peng, Borui Liu, Changzheng Wu, Yi Xie, and Guihua Yu. 2013. “Ultrathin Two-Dimensional MnO2/Graphene Hybrid Nanostructures for High-Performance, Flexible Planar Supercapacitors.” *Nano Letters* 13 (5): 2151–57. doi:10.1021/nl400600x

Peng, Weijun, Huilan Chen, Wei Wang, Yanfang Huang, and Guihong Han. 2020. "Synthesis of NiCo2S4 Nanospheres/Reduced Graphene Oxide Composite as Electrode Material for Supercapacitor." *Current Applied Physics* 20 (2): 304–9. doi:10.1016/j.cap.2019.11.023

Peng, Zhiwei, Ruquan Ye, Jason A. Mann, Dante Zakhidov, Yilun Li, Preston R. Smalley, Jian Lin, and James M. Tour. 2015. "Flexible Boron-Doped Laser-Induced Graphene Microsupercapacitors." *ACS Nano* 9 (6): 5868–75. doi:10.1021/acsnano.5b00436

Perera, Sanjaya D., Bijal Patel, Nour Nijem, Katy Roodenko, Oliver Seitz, John P. Ferraris, Yves J. Chabal, and Kenneth J. Balkus. 2011. "Vanadium Oxide Nanowire-Carbon Nanotube Binder-Free Flexible Electrodes for Supercapacitors." *Advanced Energy Materials* 1 (5): 936–45. doi:10.1002/aenm.201100221

Poonam, Kriti Sharma, Anmol Arora, and S.K. Tripathi. 2019. "Review of Supercapacitors: Materials and Devices." *Journal of Energy Storage* 21 (October 2018): 801–25. doi:10.1016/j.est.2019.01.010

Pope, Michael A., Sibel Korkut, Christian Punckt, and Ilhan A. Aksay. 2013. "Supercapacitor Electrodes Produced through Evaporative Consolidation of Graphene Oxide-Water-Ionic Liquid Gels." *Journal of The Electrochemical Society* 160 (10): A1653–660. doi:10.1149/2.017310jes

Prasad, Gautham G., Nidheesh Shetty, Simran Thakur, Rakshitha, and K.B. Bommegowda. 2019. "Supercapacitor Technology and Its Applications: A Review." In *IOP Conference Series: Materials Science and Engineering*, 012105. doi:10.1088/1757-899X/561/1/012105

Pratiwi, Nur'aini Dian, Mita Handayani, Risa Suryana, and Osamu Nakatsuka. 2019. "Fabrication of Porous Silicon Using Photolithography and Reactive Ion Etching (RIE)." *Materials Today: Proceedings* 13: 92–96. doi:10.1016/j.matpr.2019.03.194

Presser, Volker, Christopher R. Dennison, Jonathan Campos, Kevin W. Knehr, Emin C. Kumbur, and Yury Gogotsi. 2012. "The Electrochemical Flow Capacitor: A New Concept for Rapid Energy Storage and Recovery." *Advanced Energy Materials* 2 (7): 895–902. doi:10.1002/aenm.201100768

Pu, Xiong, Weiguo Hu, and Zhong Lin Wang. 2018. "Toward Wearable Self-Charging Power Systems: The Integration of Energy-Harvesting and Storage Devices." *Small* 14 (1): 1–19. doi:10.1002/smll.201702817

Qi, Dianpeng, Yan Liu, Zhiyuan Liu, Li Zhang, and Xiaodong Chen. 2017. "Design of Architectures and Materials in In-Plane Micro-Supercapacitors: Current Status and Future Challenges." *Advanced Materials* 29 (5): 1–19. doi:10.1002/adma.201602802

Qin, Jieqiong, Jianmei Gao, Xiaoyu Shi, Junyu Chang, Yanfeng Dong, Shuanghao Zheng, Xiao Wang, Liang Feng, and Zhong Shuai Wu. 2020. "Hierarchical Ordered Dual-Mesoporous Polypyrrole/Graphene Nanosheets as Bi-Functional Active Materials for High-Performance Planar Integrated System of Micro-Supercapacitor and Gas Sensor." *Advanced Functional Materials* 30 (16): 1–9. doi:10.1002/adfm.201909756

Qin, Jieqiong, Sen Wang, Feng Zhou, Pratteek Das, Shuanghao Zheng, Chenglin Sun, Xinhe Bao, and Zhong Shuai Wu. 2019. "2D Mesoporous MnO2 Nanosheets for High-Energy Asymmetric Micro-Supercapacitors in Water-in-Salt Gel Electrolyte." *Energy Storage Materials* 18 (December 2018): 397–404. doi:10.1016/j.ensm.2018.12.022

Qin, Shanshan, Qian Zhang, Xixi Yang, Mengmeng Liu, Qijun Sun, and Zhong Lin Wang. 2018. "Hybrid Piezo/Triboelectric-Driven Self-Charging Electrochromic Supercapacitor Power Package." *Advanced Energy Materials* 8 (23): 1–9. doi:10.1002/aenm.201800069

Qin, Tianfeng, Shanglong Peng, Jiaxin Hao, Yuxiang Wen, Zilei Wang, Xuefeng Wang, Deyan He, Jiachi Zhang, Juan Hou, and Guozhong Cao. 2017. "Flexible and Wearable All-Solid-State Supercapacitors with Ultrahigh Energy Density Based on a Carbon Fiber Fabric Electrode." *Advanced Energy Materials* 7 (20): 1–10. doi:10.1002/aenm.201700409

Qin, Yuancheng, and Qiang Peng. 2012. "Ruthenium Sensitizers and Their Applications in Dye-Sensitized Solar Cells." *International Journal of Photoenergy* 2012 (Ii), Article ID 291579, 21 pages. doi:10.1155/2012/291579

Raj, R. Pavul, P. Ragupathy, and S. Mohan. 2015. "Remarkable Capacitive Behavior of a Co3O4-Polyindole Composite as Electrode Material for Supercapacitor Applications." *Journal of Materials Chemistry A* 3 (48): 24338–348. doi:10.1039/c5ta07046e

Rajesh, Murugesan, C. Justin Raj, Byung Chul Kim, Bo Bae Cho, Jang Myoun Ko, and Kook Hyun Yu. 2016. "Supercapacitive Studies on Electropolymerized Natural Organic Phosphate Doped Polypyrrole Thin Films." *Electrochimica Acta* 220: 373–83. doi:10.1016/j.electacta.2016.10.118

Ramadoss, Ananthakumar, Balasubramaniam Saravanakumar, Seung Woo Lee, Young Soo Kim, Sang Jae Kim, and Zhong Lin Wang. 2015. "Piezoelectric-Driven Self-Charging Supercapacitor Power Cell." *ACS Nano* 9 (4): 4337–45. doi:10.1021/acsnano.5b00759

Rangom, Yverick, Xiaowu Tang, and Linda F. Nazar. 2015. "Carbon Nanotube-Based Supercapacitors with Excellent Ac Line Filtering and Rate Capability via Improved Interfacial Impedance." *ACS Nano* 9 (7): 7248–55. doi:10.1021/acsnano.5b02075

Raymundo-Piñero, E., K. Kierzek, J. Machnikowski, and F. Béguin. 2006. "Relationship between the Nanoporous Texture of Activated Carbons and Their Capacitance Properties in Different Electrolytes." *Carbon* 44 (12): 2498–507. doi:10.1016/j.carbon.2006.05.022

Roldán, Silvia, Daniel Barreda, Marcos Granda, Rosa Menéndez, Ricardo Santamaría, and Clara Blanco. 2015. "An Approach to Classification and Capacitance Expressions in Electrochemical Capacitors Technology." *Physical Chemistry Chemical Physics* 17 (2): 1084–92. doi:10.1039/c4cp05124f

Saleem, Junaid, Usman Bin Shahid, Mouhammad Hijab, Hamish Mackey, and Gordon McKay. 2019. "Production and Applications of Activated Carbons as Adsorbents from Olive Stones." *Biomass Conversion and Biorefinery* 9 (4): 775–802. doi:10.1007/s13399-019-00473-7

Samantara, Aneeya K., and Satyajit Ratha. 2018a. "Components of Supercapacitor." *Materials Development for Active/Passive Components of a Supercapacitor* no. i: 11–39. doi:10.1007/978-981-10-7263-5_3

Samantara, Aneeya K., and Satyajit Ratha. 2018b. "Historical Background and Present Status of the Supercapacitors." In *Materials Development for Active/Passive Components of a Supercapacitor*, 9–10. doi:10.1007/978-981-10-7263-5_2

Santos, Alan C., Barlş Çakmak, Steve Campbell, and Nikolaj T. Zinner. 2019. "Stable Adiabatic Quantum Batteries." *Physical Review E* 100 (3): 032107. doi:10.1103/PHYSREVE.100.032107/FIGURES/4/MEDIUM

Sathiya, M., A.S. Prakash, K. Ramesha, J.M. Tarascon, and A.K. Shukla. 2011. "V2O5-Anchored Carbon Nanotubes for Enhanced Electrochemical Energy Storage." *Journal of the American Chemical Society* 133 (40): 16291–299. doi:10.1021/ja207285b

Senokos, E., V. Reguero, J. Palma, J.J. Vilatela, and Rebeca Marcilla. 2016. "Macroscopic Fibres of CNTs as Electrodes for Multifunctional Electric Double Layer Capacitors: From Quantum Capacitance to Device Performance." *Nanoscale* 8 (6): 3620–28. doi:10.1039/c5nr07697h

Senthilkumar, B., P. Thenamirtham, and R. Kalai Selvan. 2011. "Structural and Electrochemical Properties of Polythiophene." *Applied Surface Science* 257 (21): 9063–67. doi:10.1016/j.apsusc.2011.05.100

Shafey, Asmaa Mohamed El. 2020. "Green Synthesis of Metal and Metal Oxide Nanoparticles from Plant Leaf Extracts and Their Applications: A Review." *Green Processing and Synthesis* 9 (1): 304–39. doi:10.1515/gps-2020-0031

Shah, Rakesh, Xianfeng Zhang, and Saikat Talapatra. 2009. "Electrochemical Double Layer Capacitor Electrodes Using Aligned Carbon Nanotubes Grown Directly on Metals." *Nanotechnology* 20 (39): 395202. doi:10.1088/0957-4484/20/39/395202

Shan, Yan, and Lian Gao. 2007. "Formation and Characterization of Multi-Walled Carbon Nanotubes/Co3O4 Nanocomposites for Supercapacitors." *Materials Chemistry and Physics* 103 (2–3): 206–10. doi:10.1016/j.matchemphys.2007.02.038

Shao, Yuanlong, Maher F. El-Kady, Jingyu Sun, Yaogang Li, Qinghong Zhang, Meifang Zhu, Hongzhi Wang, Bruce Dunn, and Richard B. Kaner. 2018. "Design and Mechanisms of Asymmetric Supercapacitors." *Chemical Reviews* 118 (18): 9233–80. doi:10.1021/acs.chemrev.8b00252

Shen, Caiwei, Xiaohong Wang, Wenfeng Zhang, and Feiyu Kang. 2011. "A High-Performance Three-Dimensional Micro Supercapacitor Based on Self-Supporting Composite Materials." *Journal of Power Sources* 196 (23): 10465–471. doi:10.1016/j.jpowsour.2011.08.007

Shi, Xiaoyu, Feng Zhou, Jiaxi Peng, Ren'an Wu, Zhong Shuai Wu, and Xinhe Bao. 2019. "One-Step Scalable Fabrication of Graphene-Integrated Micro-Supercapacitors with Remarkable Flexibility and Exceptional Performance Uniformity." *Advanced Functional Materials* 29 (50): 1–9. doi:10.1002/adfm.201902860

Shi, Zijun, Juyin Liu, Yanfang Gao, Lijun Li, and Zhenzhu Cao. 2020. "Boosted Charge Transfer in P-n Heterojunctions LaMoO3/GQDs Negative Electrode for All-Solid-State Asymmetric Supercapacitor." *Applied Surface Science* 532 (December): 147384. doi:10.1016/J.APSUSC.2020.147384

Shinde, Dhanraj B., and Vijayamohanan K. Pillai. 2013. "Electrochemical Resolution of Multiple Redox Events for Graphene Quantum Dots." *Angewandte Chemie - International Edition* 52 (9): 2482–85. doi:10.1002/anie.201208904

Shrestha, Lok Kumar, Rekha Goswami Shrestha, Subrata Maji, Bhadra P. Pokharel, Rinita Rajbhandari, Ram Lal Shrestha, Raja Ram Pradhananga, Jonathan P. Hill, and Katsuhiko Ariga. 2020. "High Surface Area Nanoporous Graphitic Carbon Materials Derived from Lapsi Seed with Enhanced Supercapacitance." *Nanomaterials* 10 (4): 728–42. doi:10.3390/nano10040728

Silva, Débora A.C. Da, Antenor J. Paulista Neto, Aline M. Pascon, Eudes E. Fileti, Leonardo R.C. Fonseca, and Hudson G. Zanin. 2021. "Combined Density Functional Theory and Molecular Dynamics Simulations to Investigate the Effects of Quantum and Double-Layer Capacitances in Functionalized Graphene as the Electrode Material of Aqueous-Based Supercapacitors." *Journal of Physical Chemistry C* 125 (10): 5518–24. doi:10.1021/acs.jpcc.0c11602

Simon, P., Y. Gogotsi, and B. Dunn. 2014. "Where Do Batteries End and Supercapacitors Begin ?" *Science* 343 (March): 1210–11.

Simon, Patrice, and Yury Gogotsi. 2008. "Materials for Electrochemical Capacitors." *Nature Materials* 7 (11): 845–54. doi:10.1038/nmat2297

Simon, Patrice, Yury Gogotsi, and Bruce Dunn. 2014. "Where Do Batteries End and Supercapacitors Begin?" *Science* 343 (6176): 1210–11. doi:10.1126/science.1249625

Singh, Adarsh Pal, Nitesh Kumar Tiwari, P.B. Karandikar, and Aman Dubey. 2015. "Effect of Electrode Shape on the Parameters of Supercapacitor." *2015 International Conference on Industrial Instrumentation and Control, ICIC 2015*, no. Icic: 669–73. doi:10.1109/IIC.2015.7150826

Singh, Madhusudan, Hanna M. Haverinen, Parul Dhagat, and Ghassan E. Jabbour. 2010. "Inkjet Printing-Process and Its Applications." *Advanced Materials* 22 (6): 673–85. doi:10.1002/adma.200901141

Sivakkumar, S.R., Wan Ju Kim, Ji Ae Choi, Douglas R. MacFarlane, Maria Forsyth, and Dong Won Kim. 2007. "Electrochemical Performance of Polyaniline Nanofibres and Polyaniline/Multi-Walled Carbon Nanotube Composite as an Electrode Material for Aqueous Redox Supercapacitors." *Journal of Power Sources* 171 (2): 1062–68. doi:10.1016/j.jpowsour.2007.05.103

Smithyman, Jesse, and Richard Liang. 2014. "Flexible Supercapacitor Yarns with Coaxial Carbon Nanotube Network Electrodes." *Materials Science and Engineering B: Solid-State Materials for Advanced Technology* 184 (1): 34–43. doi:10.1016/j.mseb.2014.01.013

Snaith, Henry J., Adam J. Moule, Cédric Klein, Klaus Meerholz, Richard H. Friend, and Michael Grätzel. 2007. "Efficiency Enhancements in Solid-State Hybrid Solar Cells via Reduced Charge Recombination and Increased Light Capture." *Nano Letters* 7 (11): 3372–76. doi:10.1021/nl071656u

Snook, Graeme A., Pon Kao, and Adam S. Best. 2011. "Conducting-Polymer-Based Supercapacitor Devices and Electrodes." *Journal of Power Sources* 196 (1): 1–12. doi:10.1016/j.jpowsour.2010.06.084

Sodano, Henry A., Gyuhae Park, Donald J. Leo, and Daniel J. Inman. 2017. "Imece2003-4 3250." *Smart Structures and Materials 2003: Smart Sensor Technology and Measurement Systems* 1–10. https://doi.org/10.1117/12.484247

Song, Ruobing, Huanyu Jin, Xing Li, Linfeng Fei, Yuda Zhao, Haitao Huang, Helen Lai-Wa Chan, Yu Wang, and Yang Chai. 2015. "A Rectification-Free Piezo-Supercapacitor with a Polyvinylidene Fluoride Separator and Functionalized Carbon Cloth Electrodes." *Journal of Materials Chemistry A* 3 (29): 14963–970. doi:10.1039/c5ta03349g

Song, Yu, Haotian Chen, Xuexian Chen, Hanxiang Wu, Hang Guo, Xiaoliang Cheng, Bo Meng, and Haixia Zhang. 2018. "All-in-One Piezoresistive-Sensing Patch Integrated with Micro-Supercapacitor." *Nano Energy* 53 (February): 189–97. doi:10.1016/j.nanoen.2018.08.041

Soni, Roby, and Sreekumar Kurungot. 2019. "Fe2P4O12-Carbon Composite as a Highly Stable Electrode Material for Electrochemical Capacitors." *New Journal of Chemistry* 43 (1): 399–406. doi:10.1039/c8nj04671a

Stern, Otto. 1924. "Zur Theorie Der Elektrolytischen Doppelschicht." *Zeitschrift Für Elektrochemie Und Angewandte Physikalische Chemie* 30 (21–22): 508–16. doi:10.1002/BBPC.192400182

Stoller, Meryl D., and Rodney S. Ruoff. 2010. "Best Practice Methods for Determining an Electrode Material's Performance for Ultracapacitors." *Energy and Environmental Science* 3 (9): 1294–301. doi:10.1039/c0ee00074d

Sun, Gengzhi, Juqing Liu, Xiao Zhang, Xuewan Wang, Hai Li, Yang Yu, Wei Huang, Hua Zhang, and Peng Chen. 2014. "Fabrication of Ultralong Hybrid Microfibers from Nanosheets of Reduced Graphene Oxide and Transition-Metal Dichalcogenides and Their Application as Supercapacitors." *Angewandte Chemie* 126 (46): 12784–788. doi:10.1002/ange.201405325

Sun, Hao, Xiao You, Yishu Jiang, Guozhen Guan, Xin Fang, Jue Deng, Peining Chen, Yongfeng Luo, and Huisheng Peng. 2014. "Self-Healable Electrically Conducting Wires for Wearable Microelectronics." *Angewandte Chemie - International Edition* 53 (36): 9526–31. doi:10.1002/anie.201405145

Sun, Leimeng, Xinghui Wang, Kang Zhang, Jianping Zou, and Qing Zhang. 2016. "Metal-Free SWNT/Carbon/MnO2 Hybrid Electrode for High Performance Coplanar Micro-Supercapacitors." *Nano Energy* 22: 11–18. doi:10.1016/j.nanoen.2015.12.007

Sung, Joo Hwan, Se Joon Kim, Soo Hwan Jeong, Eun Ha Kim, and Kun Hong Lee. 2006. "Flexible Micro-Supercapacitors." *Journal of Power Sources* 162 (2 SPEC. ISS.): 1467–70. doi:10.1016/j.jpowsour.2006.07.073

Sung, Joo Hwan, Se Joon Kim, and Kun Hong Lee. 2003. "Fabrication of Microcapacitors Using Conducting Polymer Microelectrodes." *Journal of Power Sources* 124 (1): 343–50. doi:10.1016/S0378-7753(03)00669-4

Sung, Joo Hwan, Se Joon Kim, and Kun Hong Lee. 2004. "Fabrication of All-Solid-State Electrochemical Microcapacitors." *Journal of Power Sources* 133 (2): 312–19. doi:10.1016/j.jpowsour.2004.02.003

Sunil, Vaishak, Bhupender Pal, Izan Izwan Misnon, and Rajan Jose. 2020. "Characterization of Supercapacitive Charge Storage Device Using Electrochemical Impedance Spectroscopy." *Materials Today: Proceedings* 46 (xxxx): 1588–94. doi:10.1016/j.matpr.2020.07.248

Syarif, Nirwan, Ivandini Tribidasari, and Widayanti Wibowo. 2012. "Direct Synthesis Carbon/Metal Oxide Composites for Electrochemical Capacitors Electrode." *International Transaction Journal of Engineering, Management, & Applied Sciences & Technologies* 3 (1): 21–34.

Tamilarasan, P., Ashish Kumar Mishra, and Sundara Ramaprabhu. 2011. "Graphene/Ionic Liquid Binary Electrode Material for High Performance Supercapacitor." *2011 International Conference on Nanoscience, Technology and Societal Implications, NSTSI11*, no. Cv. IEEE. doi:10.1109/NSTSI.2011.6111793

Tan, Adrian Wei Yee, Wen Sun, Ayan Bhowmik, Jun Yan Lek, Iulian Marinescu, Feng Li, Nay Win Khun, Zhili Dong, and Erjia Liu. 2018. "Effect of Coating Thickness on Microstructure, Mechanical Properties and Fracture Behaviour of Cold Sprayed Ti6Al4V Coatings on Ti6Al4V Substrates." *Surface and Coatings Technology* 349: 303–17. doi:10.1016/j.surfcoat.2018.05.060

Tao, Ran, Lin Li, Li-Jun Zhu, Yue-Dong Yan, Lin-Hai Guo, Xiao-Dong Fan, and Chang-Gan Zeng. 2020. "Giant-Capacitance-Induced Wide Quantum Hall Plateaus in Graphene on LaAlO3/SrTiO3 Heterostructures." *Chinese Physics Letters* 37 (7): 077301. doi:10.1088/0256-307X/37/7/077301

Tian, Bozhi, Xiaolin Zheng, Thomas J. Kempa, Ying Fang, Nanfang Yu, Guihua Yu, Jinlin Huang, and Charles M. Lieber. 2007. "Coaxial Silicon Nanowires as Solar Cells and Nanoelectronic Power Sources." *Nature* 449 (7164): 885–89. doi:10.1038/nature06181

Tian, Xiaocong, Mengzhu Shi, Xu Xu, Mengyu Yan, Lin Xu, Aamir Minhas-Khan, Chunhua Han, Liang He, and Liqiang Mai. 2015. "Arbitrary Shape Engineerable Spiral Micropseudocapacitors with Ultrahigh Energy and Power Densities." *Advanced Materials* 27 (45): 7476–82. doi:10.1002/adma.201503567

Tian, Xiaocong, Bei Xiao, Xu Xu, Lin Xu, Zehua Liu, Zhaoyang Wang, Mengyu Yan, Qiulong Wei, and Liqiang Mai. 2016. "Vertically Stacked Holey Graphene/Polyaniline Heterostructures with Enhanced Energy Storage for on-Chip Micro-Supercapacitors." *Nano Research* 9 (4): 1012–21. doi:10.1007/s12274-016-0989-x

Tung, Tran Thanh, Jeongha Yoo, Faisal K. Alotaibi, Md J. Nine, Ramesh Karunagaran, Melinda Krebsz, Giang T. Nguyen, Diana N.H. Tran, Jean Francois Feller, and Dusan Losic. 2016. "Graphene Oxide-Assisted Liquid Phase Exfoliation of Graphite into Graphene for Highly Conductive Film and Electromechanical Sensors." *ACS Applied Materials and Interfaces* 8 (25): 16521–532. doi:10.1021/acsami.6b04872

Uke, Santosh J., Vijay P. Akhare, Devidas R. Bambole, Anjali B. Bodade, and Gajanan N. Chaudhari. 2017. "Recent Advancements in the Cobalt Oxides, Manganese Oxides, and Their Composite as an Electrode Material for Supercapacitor: A Review." *Frontiers in Materials* 4 (August): 2–7. doi:10.3389/fmats.2017.00021

Uono, Hiroyuki, Bong-Chull Kim, Tooru Fuse, Makoto Ue, and Jun-ichi Yamaki. 2006. "Optimized Structure of Silicon/Carbon/Graphite Composites as an Anode Material for Li-Ion Batteries." *Journal of The Electrochemical Society* 153 (9): A1708. doi:10.1149/1.2218163

Wan, Houzhao, Jianjun Jiang, Jingwen Yu, Kui Xu, Ling Miao, Li Zhang, Haichao Chen, and Yunjun Ruan. 2013. "NiCo2S4 Porous Nanotubes Synthesis via Sacrificial Templates: High-Performance Electrode Materials of Supercapacitors." *CrystEngComm* 15 (38): 7649–51. doi:10.1039/c3ce41243a

Wang, Bo, Tingting Ruan, Yong Chen, Fan Jin, Li Peng, Yu Zhou, Dianlong Wang, and Shixue Dou. 2020. "Graphene-Based Composites for Electrochemical Energy Storage." *Energy Storage Materials* 24: 22–51. doi:10.1016/j.ensm.2019.08.004

Wang, Hua, Bowen Zhu, Wencao Jiang, Yun Yang, Wan Ru Leow, Hong Wang, and Xiaodong Chen. 2014. "A Mechanically and Electrically Self-Healing Supercapacitor." *Advanced Materials* 26 (22): 3638–43. doi:10.1002/adma.201305682

Wang, Huanwen, Huan Yi, Xiao Chen, and Xuefeng Wang. 2014. "Asymmetric Supercapacitors Based on Nano-Architectured Nickel Oxide/Graphene Foam and Hierarchical Porous Nitrogen-Doped Carbon Nanotubes with Ultrahigh-Rate Performance." *Journal of Materials Chemistry A* 2 (9): 3223–30. doi:10.1039/c3ta15046a

Wang, Jian Gan, Feiyu Kang, and Bingqing Wei. 2015. "Engineering of MnO2-Based Nanocomposites for High-Performance Supercapacitors." *Progress in Materials Science* 74: 51–124. doi:10.1016/j.pmatsci.2015.04.003

Wang, Jianjian, Shien Ping Feng, Yuan Yang, Nga Yu Hau, Mary Munro, Emerald Ferreira-Yang, and Gang Chen. 2015. "Thermal Charging Phenomenon in Electrical Double Layer Capacitors." *Nano Letters* 15 (9): 5784–90. doi:10.1021/acs.nanolett.5b01761

Wang, John, Julien Polleux, James Lim, and Bruce Dunn. 2007. "Pseudocapacitive Contributions to Electrochemical Energy Storage in TiO 2 (Anatase) Nanoparticles." *Journal of Physical Chemistry C* 111 (40): 14925–931. doi:10.1021/jp074464w

Wang, Lu, Xiao Feng, Lantian Ren, Qiuhan Piao, Jieqiang Zhong, Yuanbo Wang, Haiwei Li, Yifa Chen, and Bo Wang. 2015. "Flexible Solid-State Supercapacitor Based on a Metal-Organic Framework Interwoven by Electrochemically-Deposited PANI." *Journal of the American Chemical Society* 137 (15): 4920–23. doi:10.1021/jacs.5b01613

Wang, Pengfei, Yuxing Xu, Hui Liu, Yunfa Chen, Jun Yang, and Qiangqiang Tan. 2015. "Carbon/Carbon Nanotube-Supported RuO2 Nanoparticles with a Hollow Interior as Excellent Electrode Materials for Supercapacitors." *Nano Energy* 15: 116–24. doi:10.1016/j.nanoen.2015.04.006

Wang, Ronghua, Meng Han, Qiannan Zhao, Zonglin Ren, Xiaolong Guo, Chaohe Xu, Ning Hu, and Li Lu. 2017. "Hydrothermal Synthesis of Nanostructured Graphene/Polyaniline Composites as High-Capacitance Electrode Materials for Supercapacitors." *Scientific Reports* 7 (174): 1–9. doi:10.1038/srep44562

Wang, Rui, Minjie Yao, and Zhiqiang Niu. 2020. "Smart Supercapacitors from Materials to Devices." *InfoMat* 2 (1): 113–25. doi:10.1002/inf2.12037

Wang, Xianfu, Bin Liu, Rong Liu, Qiufan Wang, Xiaojuan Hou, Di Chen, Rongming Wang, and Guozhen Shen. 2014. "Fiber-Based Flexible All-Solid-State Asymmetric Supercapacitors for Integrated Photodetecting System." *Angewandte Chemie* 126 (7): 1880–84. doi:10.1002/ange.201307581

Wang, Xiaofeng, Yajiang Yin, Xiangyu Li, and Zheng You. 2014. "Fabrication of a Symmetric Micro Supercapacitor Based on Tubular Ruthenium Oxide on Silicon 3D Microstructures." *Journal of Power Sources* 252: 64–72. doi:10.1016/j.jpowsour.2013.11.109

Wang, Xinyu, Qiongqiong Lu, Chen Chen, Mo Han, Qingrong Wang, Haixia Li, Zhiqiang Niu, and Jun Chen. 2017. "A Consecutive Spray Printing Strategy to Construct and Integrate Diverse Supercapacitors on Various Substrates." *ACS Applied Materials and Interfaces* 9 (34): 28612–619. doi:10.1021/acsami.7b08833

Wang, Yaling, Yan Zhang, Guolong Wang, Xiaowei Shi, Yide Qiao, Jiamei Liu, Heguang Liu, Anandha Ganesh, and Lei Li. 2020. "Direct Graphene-Carbon Nanotube Composite Ink Writing All-Solid-State Flexible Microsupercapacitors with High Areal Energy Density." *Advanced Functional Materials* 30 (16): 1–9. doi:10.1002/adfm.201907284

Wang, Zhikui, and Qinmin Pan. 2017. "An Omni-Healable Supercapacitor Integrated in Dynamically Cross-Linked Polymer Networks." *Advanced Functional Materials* 27 (24): 1–8. doi:10.1002/adfm.201700690

Wang, Zhong Lin. 2010. "Toward Self-Powered Sensor Networks." *Nano Today* 5 (6): 512–14. doi:10.1016/j.nantod.2010.09.001

Wang, Zhong Lin, and Jinhui Song. 2006. "Piezoelectric Nanogenerators Based on Zinc Oxide Nanowire Arrays." *Science* 312 (5771): 242–46. doi:10.1126/science.1124005

Wei, Huige, Dapeng Cui, Junhui Ma, Liqiang Chu, Xiaoyu Zhao, Haixiang Song, Hu Liu, Tao Liu, Ning Wang, and Zhanhu Guo. 2017. "Energy Conversion Technologies towards Self-Powered Electrochemical Energy Storage Systems: The State of the Art and Perspectives." *Journal of Materials Chemistry A* 5 (5): 1873–94. doi:10.1039/C6TA09726J

Wei, Z.G., R. Sandstrorö m, and S. Miyazaki. 1998. "Review Shape-Memory Meterials and Hybrid Composites for Smart Systems." *Journal of Materials Science* 33 (15): 3743–62.

Weng, Zhe, Yang Su, Da Wei Wang, Feng Li, Jinhong Du, and Hui Ming Cheng. 2011. "Graphene-Cellulose Paper Flexible Supercapacitors." *Advanced Energy Materials* 1 (5): 917–22. doi:10.1002/aenm.201100312

Winter, Martin, and Ralph J. Brodd. 2004. "What Are Batteries, Fuel Cells, and Supercapacitors?" *Chemical Reviews* 104 (10): 4245–69. doi:10.1021/cr020730k

Wu, Hong Ying, and Huan Wen Wang. 2012. "Electrochemical Synthesis of Nickel Oxide Nanoparticulate Films on Nickel Foils for High-Performance Electrode Materials of Supercapacitors." *International Journal of Electrochemical Science* 7 (5): 4405–17.

Wu, Zhen Kun, Ziyin Lin, Liyi Li, Bo Song, Kyoung sik Moon, Shu Lin Bai, and Ching Ping Wong. 2014. "Flexible Micro-Supercapacitor Based on in-Situ Assembled Graphene on Metal Template at Room Temperature." *Nano Energy* 10: 222–28. doi:10.1016/j.nanoen.2014.09.019

Wu, Zhenkun, Liyi Li, Ziyin Lin, Bo Song, Zhuo Li, Kyoung Sik Moon, Ching Ping Wong, and Shu Lin Bai. 2015. "Alternating Current Line-Filter Based on Electrochemical Capacitor Utilizing Template-Patterned Graphene." *Scientific Reports* 5 (June): 1–7. doi:10.1038/srep10983

Wu, Zhong, Lin Li, Jun Min Yan, and Xin Bo Zhang. 2017. "Materials Design and System Construction for Conventional and New-Concept Supercapacitors." *Advanced Science* 4 (6): 1600382. doi:10.1002/ADVS.201600382

Wu, Zhong Shuai, Khaled Parvez, Xinliang Feng, and Klaus Müllen. 2013. "Graphene-Based in-Plane Micro-Supercapacitors with High Power and Energy Densities." *Nature Communications* 4: 1–8. doi:10.1038/ncomms3487

Wu, Zhong Shuai, Khaled Parvez, Andreas Winter, Henning Vieker, Xianjie Liu, Sheng Han, Andrey Turchanin, Xinliang Feng, and Klaus Müllen. 2014. "Layer-by-Layer Assembled Heteroatom-Doped Graphene Films with Ultrahigh Volumetric Capacitance and Rate Capability for Micro-Supercapacitors." *Advanced Materials* 26 (26): 4552–58. doi:10.1002/adma.201401228

Wu, Zhong Shuai, Wencai Ren, Da Wei Wang, Feng Li, Bilu Liu, and Hui Ming Cheng. 2010. "High-Energy MnO2 Nanowire/Graphene and Graphene Asymmetric Electrochemical Capacitors." *ACS Nano* 4 (10): 5835–42. doi:10.1021/nn101754k

Wu, Zhong Shuai, Guangmin Zhou, Li Chang Yin, Wencai Ren, Feng Li, and Hui Ming Cheng. 2012. "Graphene/Metal Oxide Composite Electrode Materials for Energy Storage." *Nano Energy* 1 (1): 107–31. doi:10.1016/j.nanoen.2011.11.001

Xiang, Chengcheng, Ming Li, Mingjia Zhi, Ayyakkannu Manivannan, and Nianqiang Wu. 2013. "A Reduced Graphene Oxide/Co3O4 Composite for Supercapacitor Electrode." *Journal of Power Sources* 226: 65–70. doi:10.1016/j.jpowsour.2012.10.064

Xiao, Xu, Longyan Yuan, Junwen Zhong, Tianpeng Ding, Yu Liu, Zhixiang Cai, Yaoguang Rong, Hongwei Han, Jun Zhou, and Zhong Lin Wang. 2011. "High-Strain Sensors Based on ZnO Nanowire/Polystyrene Hybridized Flexible Films." *Advanced Materials* 23 (45): 5440–44. doi:10.1002/adma.201103406

Xie, Keyu, and Bingqing Wei. 2014. "Materials and Structures for Stretchable Energy Storage and Conversion Devices." *Advanced Materials* 26 (22): 3592–617. doi:10.1002/adma.201305919

Xing, Lili, Yuxin Nie, Xinyu Xue, and Yan Zhang. 2014. "PVDF Mesoporous Nanostructures as the Piezo-Separator for a Self-Charging Power Cell." *Nano Energy* 10: 44–52. doi:10.1016/j.nanoen.2014.09.004

Xiong, Guoping, Pingge He, Lei Liu, Tengfei Chen, and Timothy S. Fisher. 2015. "Plasma-Grown Graphene Petals Templating Ni-Co-Mn Hydroxide Nanoneedles for High-Rate and Long-Cycle-Life Pseudocapacitive Electrodes." *Journal of Materials Chemistry A* 3 (45): 22940–948. doi:10.1039/c5ta05441a

Xiong, Guoping, Chuizhou Meng, Ronald G. Reifenberger, Pedro P. Irazoqui, and Timothy S. Fisher. 2014. "A Review of Graphene-Based Electrochemical Microsupercapacitors." *Electroanalysis* 26 (1): 30–51. doi:10.1002/elan.201300238

Xu, Jing, Hui Wu, Linfeng Lu, Siu Fung Leung, Di Chen, Xiaoyuan Chen, Zhiyong Fan, Guozhen Shen, and Dongdong Li. 2014. "Integrated Photo-Supercapacitor Based on Bi-Polar TiO2 Nanotube Arrays with Selective One-Side Plasma-Assisted Hydrogenation." *Advanced Functional Materials* 24 (13): 1840–46. doi:10.1002/adfm.201303042

Xu, Kai, Zheng Hang Sun, Wuxin Liu, Yu Ran Zhang, Hekang Li, Hang Dong, Wenhui Ren, et al. 2020. "Probing Dynamical Phase Transitions with a Superconducting Quantum Simulator." *Science Advances* 6 (25): 4935–52. doi:10.1126/SCIADV.ABA4935/SUPPL_FILE/ABA4935_SM.PDF

Xu, Kui, Hui Shao, Zifeng Lin, Céline Merlet, Guang Feng, Jixin Zhu, and Patrice Simon. 2020. "Computational Insights into Charge Storage Mechanisms of Supercapacitors." *Energy and Environmental Materials* 3 (3): 235–46. doi:10.1002/eem2.12124

Xu, Sheng, Benjamin J. Hansen, and Zhong Lin Wang. 2010. "Piezoelectric-Nanowire-Enabled Power Source for Driving Wireless Microelectronics." *Nature Communications* 1 (7): 1–5. doi:10.1038/ncomms1098

Xue, Xinyu, Ping Deng, Bin He, Yuxin Nie, Lili Xing, Yan Zhang, and Zhong Lin Wang. 2014. "Flexible Self-Charging Power Cell for One-Step Energy Conversion and Storage." *Advanced Energy Materials* 4 (5): 1–5. doi:10.1002/aenm.201301329

Xue, Xinyu, Sihong Wang, Wenxi Guo, Yan Zhang, and Zhong Lin Wang. 2012. "Hybridizing Energy Conversion and Storage in a Mechanical-to- Electrochemical Process for Self-Charging Power Cell." *Nano Letters* 12 (9): 5048–54. doi:10.1021/nl302879t

Yan, Jun, Junpeng Liu, Zhuangjun Fan, Tong Wei, and Lijun Zhang. 2012. "High-Performance Supercapacitor Electrodes Based on Highly Corrugated Graphene Sheets." *Carbon* 50 (6): 2179–88. doi:10.1016/j.carbon.2012.01.028

Yan, Jun, Tong Wei, Zhuangjun Fan, Weizhong Qian, Milin Zhang, Xiande Shen, and Fei Wei. 2010. "Preparation of Graphene Nanosheet/Carbon Nanotube/Polyaniline Composite as Electrode Material for Supercapacitors." *Journal of Power Sources* 195 (9): 3041–45. doi:10.1016/j.jpowsour.2009.11.028

Yan, Xingbin, Zhixin Tai, Jiangtao Chen, and Qunji Xue. 2011. "Fabrication of Carbon Nanofiber-Polyaniline Composite Flexible Paper for Supercapacitor." *Nanoscale* 3 (1): 212–16. doi:10.1039/c0nr00470g

Yang, Canhui, and Zhigang Suo. 2018. "Hydrogel Ionotronics." *Nature Reviews Materials 2018 3:6* 3 (6): 125–42. doi:10.1038/s41578-018-0018-7

Yang, Ching Hua, Yu Cheng Hsiao, and Lu Yin Lin. 2021. "Novel in Situ Synthesis of Freestanding Carbonized ZIF67/Polymer Nanofiber Electrodes for Supercapacitors via Electrospinning and Pyrolysis Techniques." *ACS Applied Materials and Interfaces* 13 (35): 41637–648. doi:10.1021/acsami.1c10985

Yang, Hao, Santhakumar Kannappan, Amaresh S. Pandian, Jae Hyung Jang, Yun Sung Lee, and Wu Lu. 2017. "Graphene Supercapacitor with Both High Power and Energy Density." *Nanotechnology* 28 (44): 445401–11. doi:10.1088/1361-6528/aa8948

Yang, Peihua, Yuzhi Li, Ziyin Lin, Yong Ding, Song Yue, Ching Ping Wong, Xiang Cai, Shaozao Tan, and Wenjie Mai. 2014. "Worm-like Amorphous MnO2 Nanowires Grown on Textiles for High-Performance Flexible Supercapacitors." *Journal of Materials Chemistry A* 2 (3): 595–99. doi:10.1039/c3ta14275b

Yang, Peihua, and Wenjie Mai. 2015. "Large-Scale Fabrication of Pseudocapacitive Glass Windows That Combine Electrochromism and Energy Storage." *Photonics for Energy, PFE* 2015: 12129–133. doi:10.1002/ange.201407365

Yang, Qin, Siu Kwong Pang, and Kam Chuen Yung. 2014. "Study of PEDOT-PSS in Carbon Nanotube/Conducting Polymer Composites as Supercapacitor Electrodes in Aqueous Solution." *Journal of Electroanalytical Chemistry* 728: 140–47. doi:10.1016/j.jelechem.2014.06.033

Yang, Qinghao, Zhenzhong Hou, and Tianzhu Huang. 2015. "Self-Assembled Polypyrrole Film by Interfacial Polymerization for Supercapacitor Applications." *Journal of Applied Polymer Science* 132 (11): 4–8. doi:10.1002/app.41615

Yang, Xiaowei, Chi Cheng, Yufei Wang, Ling Qiu, and Dan Li. 2013. "Liquid-Mediated Dense Integration of Graphene Materials for Compact Capacitive Energy Storage." *Science* 341 (6145): 534–37. doi:10.1126/science.1239089

Yang, Yu, Qiyao Huang, Liyong Niu, Dongrui Wang, Casey Yan, Yiyi She, and Zijian Zheng. 2017. "Waterproof, Ultrahigh Areal-Capacitance, Wearable Supercapacitor Fabrics." *Advanced Materials* 29 (19): 160669–78. doi:10.1002/ADMA.201606679

Yang, Yu, Yunlong Xi, Junzhi Li, Guodong Wei, N.I. Klyui, and Wei Han. 2017. "Flexible Supercapacitors Based on Polyaniline Arrays Coated Graphene Aerogel Electrodes." *Nanoscale Research Letters* 12: 1–9. doi:10.1186/s11671-017-2159-9

Yang, Yun, Dandan Yu, Hua Wang, and Lin Guo. 2017. "Smart Electrochemical Energy Storage Devices with Self-Protection and Self-Adaptation Abilities." *Advanced Materials* 29 (45): 1–13. doi:10.1002/adma.201703040

Yang, Zhe, Hongdan Peng, Weizhi Wang, and Tianxi Liu. 2010. "Crystallization Behavior of Poly(ε-Caprolactone)/Layered Double Hydroxide Nanocomposites." *Journal of Applied Polymer Science* 116 (5): 2658–67. doi:10.1002/app

Yang, Zhibin, Jue Deng, Xuli Chen, Jing Ren, and Huisheng Peng. 2013. "A Highly Stretchable, Fiber-Shaped Supercapacitor." *Angewandte Chemie* 125 (50): 13695–699. doi:10.1002/ange.201307619

Yaseen, Muhammad, Muhammad Arif Khan Khattak, Muhammad Humayun, Muhammad Usman, Syed Shaheen Shah, Shaista Bibi, Bakhtiar Syed Ul Hasnain, et al. 2021. "A Review of Supercapacitors: Materials Design, Modification, and Applications." *Energies* 14 (22): 1–40. doi:10.3390/en14227779

Ye, Jianglin, Huabing Tan, Shuilin Wu, Kun Ni, Fei Pan, Jie Liu, Zhuchen Tao, et al. 2018. "Direct Laser Writing of Graphene Made from Chemical Vapor Deposition for Flexible, Integratable Micro-Supercapacitors with Ultrahigh Power Output." *Advanced Materials* 30 (27): 1–8. doi:10.1002/adma.201801384

Ye, Ruquan, Dustin K. James, and James M. Tour. 2018. "Laser-Induced Graphene." *Accounts of Chemical Research* 51 (7): 1609–20. doi:10.1021/acs.accounts.8b00084

Yoo, Hyun Deog, Yifei Li, Yanliang Liang, Yucheng Lan, Feng Wang, and Yan Yao. 2016. "Intercalation Pseudocapacitance of Exfoliated Molybdenum Disulfide for Ultrafast Energy Storage." *ChemNanoMat* 2 (7): 688–91. doi:10.1002/cnma.201600117

Yoo, Jung Joon, Kaushik Balakrishnan, Jingsong Huang, Vincent Meunier, Bobby G. Sumpter, Anchal Srivastava, Michelle Conway, et al. 2011. "Ultrathin Planar Graphene Supercapacitors." *Nano Letters* 11 (4): 1423–27. doi:10.1021/nl200225j

Yoo, Yongju, Min Seop Kim, Jong Kook Kim, Yong Sin Kim, and Woong Kim. 2016. "Fast-Response Supercapacitors with Graphitic Ordered Mesoporous Carbons and Carbon Nanotubes for AC Line Filtering." *Journal of Materials Chemistry A* 4 (14): 5062–68. doi:10.1039/c6ta00921b

Yu, Aiping, Victor Chabot, and Jiujun Zhang. 2013. *Electrochemical Supercapacitors for Energy Storage and Delivery*. Taylor & Francis.

Yu, Lianghao, Zhaodi Fan, Yuanlong Shao, Zhengnan Tian, Jingyu Sun, and Zhongfan Liu. 2019. "Versatile N-Doped MXene Ink for Printed Electrochemical Energy Storage Application." *Advanced Energy Materials* 9 (34): 1–8. doi:10.1002/aenm.201901839

Yu, Lili, Yi Hsien Lee, Xi Ling, Elton J.G. Santos, Yong Cheol Shin, Yuxuan Lin, Madan Dubey, et al. 2014. "Graphene/MoS2 Hybrid Technology for Large-Scale Two-Dimensional Electronics." *Nano Letters* 14 (6): 3055–63. doi:10.1021/NL404795Z/SUPPL_FILE/NL404795Z_SI_001.PDF

Yu, Pingping, Yingzhi Li, Xinyi Yu, Xin Zhao, Lihao Wu, and Qinghua Zhang. 2013. "Polyaniline Nanowire Arrays Aligned on Nitrogen-Doped Carbon Fabric for High-Performance Flexible Supercapacitors." *Langmuir* 29 (38): 12051–058. doi:10.1021/la402404a

Yuan, Longyan, Xi Hong Lu, Xu Xiao, Teng Zhai, Junjie Dai, Fengchao Zhang, Bin Hu, et al. 2012. "Flexible Solid-State Supercapacitors Based on Carbon Nanoparticles/MnO 2 Nanorods Hybrid Structure." *ACS Nano* 6 (1): 656–61. doi:10.1021/nn2041279

Yue, Yang, Nishuang Liu, Yanan Ma, Siliang Wang, Weijie Liu, Cheng Luo, Hang Zhang, et al. 2018. "Highly Self-Healable 3D Microsupercapacitor with MXene-Graphene Composite Aerogel." *ACS Nano* 12 (5): 4224–32. doi:10.1021/acsnano.7b07528

Zeng, H.M., Y. Zhao, Y.J. Hao, Q.Y. Lai, J.H. Huang, and X.Y. Ji. 2009. "Preparation and Capacitive Properties of Sheet V6O13 for Electrochemical Supercapacitor." *Journal of Alloys and Compounds* 477 (1–2): 800–04. doi:10.1016/j.jallcom.2008.10.100

Zeng, Ling, Xuechun Lou, Junhui Zhang, Chun Wu, Jie Liu, and Chuankun Jia. 2019. "Carbonaceous Mudstone and Lignin-Derived Activated Carbon and Its Application for Supercapacitor Electrode." *Surface and Coatings Technology* 357 (January): 580–86. doi:10.1016/J.SURFCOAT.2018.10.041

Zeng, Sha, Hongyuan Chen, Feng Cai, Yirang Kang, Minghai Chen, and Qingwen Li. 2015. "Electrochemical Fabrication of Carbon Nanotube/Polyaniline Hydrogel Film for All-Solid-State Flexible Supercapacitor with High Areal Capacitance." *Journal of Materials Chemistry A* 3 (47): 23864–870. doi:10.1039/c5ta05937b

Zhai, Shengli, Wenchao Jiang, Li Wei, H. Enis Karahan, Yang Yuan, Andrew Keong Ng, and Yuan Chen. 2015. "All-Carbon Solid-State Yarn Supercapacitors from Activated Carbon and Carbon Fibers for Smart Textiles." *Materials Horizons* 2 (6). Royal Society of Chemistry: 598–605. doi:10.1039/c5mh00108k

Zhang, Fan, Tengfei Zhang, Xi Yang, Long Zhang, Kai Leng, Yi Huang, and Yongsheng Chen. 2013. "A High-Performance Supercapacitor-Battery Hybrid Energy Storage Device Based on Graphene-Enhanced Electrode Materials with Ultrahigh Energy Density." *Energy and Environmental Science* 6 (6): 1623–32. doi:10.1039/c3ee40509e

Zhang, Kai, Li Li Zhang, X.S. Zhao, and Jishan Wu. 2010. "Graphene/Polyaniline Nanofiber Composites as Supercapacitor Electrodes." *Chemistry of Materials* 22 (4): 1392–401. doi:10.1021/cm902876u

Zhang, Lei, Xiaosong Hu, Zhenpo Wang, Fengchun Sun, and David G. Dorrell. 2018. "A Review of Supercapacitor Modeling, Estimation, and Applications: A Control/Management Perspective." *Renewable and Sustainable Energy Reviews* 81 (February): 1868–78. doi:10.1016/j.rser.2017.05.283

Zhang, Li, and X.S. Zhao. 2009a. "Carbon-Based Materials as Supercapacitor Electrodes." *Chemical Society Reviews* 38 (9): 2520–31. doi:10.1039/B813846J

Zhang, Li, and X.S. Zhao. 2009b. "Carbon-Based Materials as Supercapacitor Electrodes." *Chemical Society Reviews* 38 (9): 2520–31. doi:10.1039/B813846J

Zhang, Shiguo, Ai Ikoma, Kazuhide Ueno, Zhengjian Chen, Kaoru Dokko, and Masayoshi Watanabe. 2015. "Protic-Salt-Derived Nitrogen/Sulfur-Codoped Mesoporous Carbon for the Oxygen Reduction Reaction and Supercapacitors." *ChemSusChem* 8 (9): 1608–17. doi:10.1002/cssc.201403320

Zhang, Steven L., Qiu Jiang, Zhiyi Wu, Wenbo Ding, Lei Zhang, Husam N. Alshareef, and Zhong Lin Wang. 2019. "Energy Harvesting-Storage Bracelet Incorporating Electrochemical Microsupercapacitors Self-Charged from a Single Hand Gesture." *Advanced Energy Materials* 9 (18): 1–7. doi:10.1002/aenm.201900152

Zhang, Xiong, Liyan Ji, Shichao Zhang, and Wensheng Yang. 2007. "Synthesis of a Novel Polyaniline-Intercalated Layered Manganese Oxide Nanocomposite as Electrode Material for Electrochemical Capacitor." *Journal of Power Sources* 173 (2 SPEC. ISS.): 1017–23. doi:10.1016/j.jpowsour.2007.08.083

Zhang, Yi Zhou, Yang Wang, Tao Cheng, Lan Qian Yao, Xiangchun Li, Wen Yong Lai, and Wei Huang. 2019. "Printed Supercapacitors: Materials, Printing and Applications." *Chemical Society Reviews* 48 (12): 3229–64. doi:10.1039/c7cs00819h

Zhao, Chen, Yuqing Liu, Stephen Beirne, Joselito Razal, and Jun Chen. 2018. "Recent Development of Fabricating Flexible Micro-Supercapacitors for Wearable Devices." *Advanced Materials Technologies* 3 (9): 1–16. doi:10.1002/admt.201800028

Zhao, Cuimei, and Weitao Zheng. 2015. "A Review for Aqueous Electrochemical Supercapacitors." *Frontiers in Energy Research* 3 (May): 1–11. doi:10.3389/fenrg.2015.00023

Zhao, D., H. Wang, Z.U. Khan, J.C. Chen, R. Gabrielsson, M.P. Jonsson, M. Berggren, and X. Crispin. 2016. "Ionic Thermoelectric Supercapacitors." *Energy and Environmental Science* 9 (4): 1450–57. doi:10.1039/c6ee00121a

Zhao, Xiaoyu, Yingbing Zhang, Yanfei Wang, and Huige Wei. 2019. "Battery-Type Electrode Materials for Sodium-Ion Capacitors." *Batteries and Supercaps* 2 (11): 899–917. doi:10.1002/batt.201900082

Zhao, Xin, Qinghua Zhang, Yanping Hao, Yingzhi Li, Ying Fang, and Dajun Chen. 2010. "Alternate Multilayer Films of Poly(Vinyl Alcohol) and Exfoliated Graphene Oxide Fabricated via a Facial Layer-by-Layer Assembly." *Macromolecules* 43 (22): 9411–16. doi:10.1021/ma101456y

Zheng, J.P., and T.R. Jow. 1995. "A New Charge Storage Mechanism for Electrochemical Capacitors." *Journal of The Electrochemical Society* 142 (1): L6–L8. doi:10.1149/1.2043984

Zheng, Shasha, Qing Li, Huaiguo Xue, Huan Pang, and Qiang Xu. 2020. "A Highly Alkaline-Stable Metal Oxide@metal-Organic Framework Composite for High-Performance Electrochemical Energy Storage." *National Science Review* 7 (2): 305–14. doi:10.1093/nsr/nwz137

Zhi, Mingjia, Chengcheng Xiang, Jiangtian Li, Ming Li, and Nianqiang Wu. 2013. "Nanostructured Carbon-Metal Oxide Composite Electrodes for Supercapacitors: A Review." *Nanoscale* 5 (1): 72–88. doi:10.1039/c2nr32040a

Zhong, Cheng, Yida Deng, Wenbin Hu, Jinli Qiao, Lei Zhang, and Jiujun Zhang. 2015. "A Review of Electrolyte Materials and Compositions for Electrochemical Supercapacitors." *Chemical Society Reviews* 44 (21): 7484–539. doi:10.1039/c5cs00303b

Zhu, Guang, Rusen Yang, Sihong Wang, and Zhong Lin Wang. 2010. "Flexible High-Output Nanogenerator Based on Lateral ZnO Nanowire Array." *Nano Letters* 10 (8): 3151–55. doi:10.1021/nl101973h

Zhu, Minshen, Yang Huang, Yan Huang, Hongfei Li, Zifeng Wang, Zengxia Pei, Qi Xue, Huiyuan Geng, and Chunyi Zhi. 2017. "A Highly Durable, Transferable, and Substrate-Versatile High-Performance All-Polymer Micro-Supercapacitor with Plug-and-Play Function." *Advanced Materials* 29 (16): 1–7. doi:10.1002/adma.201605137

Zhu, Yachao, Khalil Rajouâ, Steven Le Vot, Olivier Fontaine, Patrice Simon, and Frédéric Favier. 2020. "Modifications of MXene Layers for Supercapacitors." *Nano Energy* 73 (March): 104734–43. doi:10.1016/j.nanoen.2020.104734

Index